Liquid Crystals in Photovoltaics

Liquid Crystals in Photovoltaics

An Introduction

Luz J. Martínez-Miranda

CRC Press
Taylor & Francis Group
Boca Raton London New York

CRC Press is an imprint of the
Taylor & Francis Group, an **informa** business

First edition published 2022
by CRC Press
6000 Broken Sound Parkway NW, Suite 300, Boca Raton, FL 33487-2742

and by CRC Press
2 Park Square, Milton Park, Abingdon, Oxon, OX14 4RN

Library of Congress Control Number: 2021931533

ISBN: 978-0-8153-8621-6 (hbk)
ISBN: 978-1-032-04140-7 (pbk)
ISBN: 978-1-351-17578-4 (ebk)

DOI: 10.1201/9781351175784

Typeset in Minion
by codeMantra

To the memory of my parents, Luz Marina Miranda-Meléndez and José Martínez-Mateo, who encouraged me to study and to excel in science. And in memory of my maternal grandmother, Celia Meléndez-Ruiz, whose love of learning influenced my mother and myself.

Contents

Preface

ORGANIC PHOTOVOLTAICS HAVE SEEN a large increase in their cells' efficiency in the last few years. At press time, this increase should be about 74% in a year. Liquid crystals have been a main contributor to this increase. However, to this date, and the best of our ability, no book combines the basics of liquid crystals with that of organic photovoltaics. This book combines the basic knowledge of liquid crystals and photovoltaics, specifically organic photovoltaics. Both fields are multi-disciplinary. The liquid crystal field requires knowledge of basic and applied physics, chemistry, and biology. It requires knowledge of materials science and other engineering fields in the development of applications. The field of organic photovoltaics requires knowledge of physics, chemistry, and materials science in order to choose the appropriate materials. Both fields require knowledge of other engineering fields to analyze the electronic and mechanical behavior of the cells.

The use of liquid crystals in organic photovoltaics or hybrid photovoltaics has seen an increase in the last few years. Liquid crystals are used as a dopant, a component, or both of an organic photovoltaic. This book investigates the properties that make the liquid crystals ideal materials for organic photovoltaics for those investigators interested in organic photovoltaics who may not necessarily be an expert in liquid crystals. It combines traditional knowledge in liquid crystals and organic electronics with knowledge acquired through materials computation to obtain a general combined picture of the two fields. The two theoretical descriptions are compared, of liquid crystal electronics and organic electronics in general, to emphasize how adding a liquid crystal can improve on the electronic properties.

The purpose of this book is to introduce young investigators, including graduate students and advanced undergraduate students, to both liquid crystals and organic photovoltaics.

Author

Luz J. Martínez-Miranda is associate professor of materials science and engineering at the University of Maryland, College Park, MD. She is a fellow of the American Association for the Advancement of Science (AAAS) and the American Physical Society (APS), and winner of the 2013 Edward A. Bouchet Prize of the APS for her research in liquid crystals. She serves on the board of the International Liquid Crystal Society (ILCS). She served as president and board member of the National Society for Hispanic Physicists (NSHP) and was a board member of the Society for the Advancement of Chicanos and Native Americans in Science (SACNAS). She has been working with liquid crystals since she was a graduate student. She has worked on the basic properties of liquid crystals, and their possible applications, both in bioapplications and in industrial applications.

CHAPTER 1

What Are Liquid Crystals and How Do We Measure Their Order?

THE MOST COMMON PHASES of matter are (a) the solid state, which is characterized either by the crystal structure or by atoms or molecules that are fixed in space, (b) the liquid state, characterized by the density and by flow of the atoms or molecules and (c) the gas phase. The atoms or molecules in a crystalline solid are ordered, both positionally and orientationally, while they are not in a liquid. Sometimes an intermediate phase occurs, as shown in the center panel of Figure 1.1. The intermediate state has structure and/or rigidity, and it can flow. The former characteristic is associated with a solid, and the latter is associated with a liquid. These phases are known as liquid crystals. It has the crystallinity or rigidity of the solid and the flowing properties of a liquid as

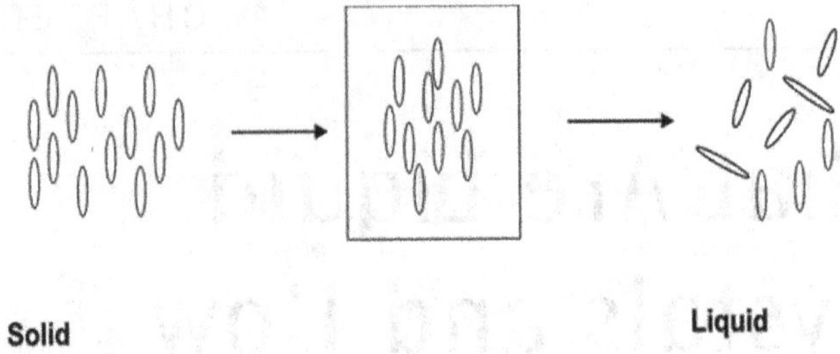

Solid   **Liquid**

FIGURE 1.1   A liquid crystal phase happening between the solid and the liquid phase of a material.

mentioned above. A liquid crystal must satisfy certain conditions to have these intermediate characteristics.

Cholesteryl benzoate (Figure 1.2a) was one of the first liquid crystals reported by Friedrich Reinitzer to have two melting points at 145°C and 178.5°C. The shape of this molecule gives the main properties needed to exhibit liquid crystalline properties.

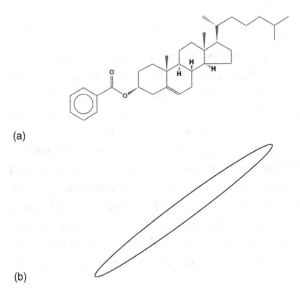

(a)

(b)

FIGURE 1.2   (a) The chemical structure of cholesteryl benzoate, used with permission of NIST; (b) the ellipsoidal shape of this molecule.

(1) It has an anisotropic shape; (2) it has rigidity in its central core, as shown by the ring structures; and (3) it has ends that are somewhat flexible and elongated, as shown by the carbon chain. These last ones can be either on one side of the molecule as is the case with cholesteryl benzoate or on both ends of the molecule. If the structure is substituted by the ellipsoid as shown in Figure 1.2b, then we have a physicist's and/or engineer's (or anybody who does not have chemist in their title) view of a liquid crystal. The anisotropic shape indicates the fact that not all directions are equivalent. Properties that relate to the structure of the molecules, including the fluid properties, will be anisotropic too. Thus, any property must be expressed as a tensor.

TYPES OF LIQUID CRYSTAL

One can find different types of liquid crystals, based on this description. We have already observed the calamitic liquid crystal, where the molecule is rod- or ellipsoidal-shaped (Figure 1.2b). Disc-shaped molecules, where one of the axes is much shorter than the other two, also satisfy the description of a liquid crystal. The molecule consists of a central core of four rings, three of which are planar (Figure 1.3) in this case. This central core is surrounded by other planar rings that have at least one long carbon chain attached to them. The calamitic and the discotic liquid crystals may exhibit polymeric liquid crystal phases. The polymeric liquid crystal phases can have the liquid crystal or mesomorphic unit as part of the polymer backbone or hanging from the polymer backbone as a side chain. The molecules that belong to these three groups are also known as thermotropic because the temperature controls where the liquid crystal phase may appear and be stabilized.

Another group of anisotropic molecules is known as lyotropic liquid crystals. These molecules need a solvent to form the liquid crystal phases. Soaps and phospholipids belong to this group. The concentration of the solvent, or of the liquid crystal in the solvent, controls the exhibited liquid crystal phase, although temperature

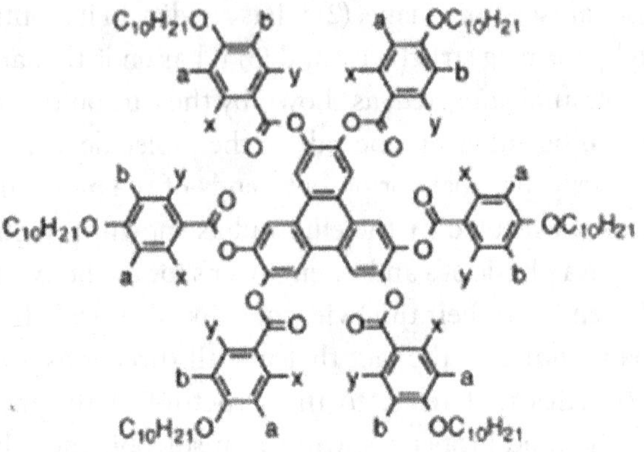

FIGURE 1.3 The general structure of a discotic liquid crystal. (Used with permission of Collings and Hird, *Introduction to Liquid Crystals; Chemistry and Physics*, 1st Edition, 1997, Taylor & Francis, Bristol, PA.)

can control the liquid crystal phase as well. DNA and Kevlar are polymer structures that belong to this last group.

The anisotropy of these liquid crystals means that the structural, electrical, optical properties and viscosity properties have different values depending on what direction we are measuring them.

LIQUID CRYSTAL PHASES

Let us consider some of the phase structures of calamitic liquid crystals. Suppose that this intermediate phase consists of the molecules all pointing along one direction but not organized in layers (Figure 1.4a). This phase will be optically uniaxial because the refractive index along one direction is different than along the other two. The long axis of the molecule is aligned along the arrow's direction. This direction is called the director, **n**. The molecules in this phase are not aligned in layers. This phase is known as the nematic, from the Greek word for thread, since it looks crisscrossed with threadlike structures under a polarizing microscope. Let us now assume that these molecules are arranged in layers (Figure 1.4b). They are still uniaxial, and the molecules point along the director. These phases feel like soap when it is rubbed

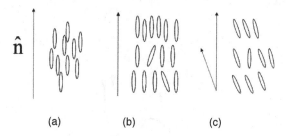

FIGURE 1.4 Three of the phases obtained for calamitic liquid crystals: (a) nematic, (b) smectic-A, and (c) smectic-C.

between the hands, and hence they are called smectic phases, for the Greek word for soap. The particular smectic phase we have described is the smectic-A phase. Another phase has the molecules tilted at an angle with respect to the director (Figure 1.4c). This phase is optically biaxial because the refractive indices along all three directions are different. If the molecules are projected in the plane perpendicular to the director (or parallel to the layers), the projection gives an ellipse that has two axes (Figure 1.5). The smectic phase has a large polymorphism, based on whether the molecules are aligned with the director or at an angle with the director. If we take into consideration the order found in one- or two- dimensions and by taking into account the polarization of the molecules, we can generate many other smectic phases.

Discotics exhibit two types of mesophases: the nematics and the columnar. There are two nematic phases: the N_D phase and the N_c phase. The N_D phase has a structure similar to the calamitic nematic, except that in this case the short axes are almost all parallel to each other and these are parallel to the director. The N_c phase consists of short columnar arrangements, where the correlation lengths are much smaller compared with the columnar phases. The polymorphism of the columnar phases depends on how they pack in the plane perpendicular to the director, whether they are hexagonal, rectangular, or oblique, and whether these packings are ordered or disordered within the columns.

The original classification of the liquid crystals was performed by Georges Friedel who subdivided the liquid crystals into three

FIGURE 1.5 The projection on the plane for a tilted phase yields an ellipse with three indices of refraction.

categories: nematic, smectics and cholesterics. The cholesterics are named after cholesterol, whose derivatives were the first to be studied for their liquid crystal characteristics, such as cholesteryl benzoate (Figure 1.2). These are chiral liquid crystals that lack inversion symmetry. For nematics, the director rotates in a helical fashion at a perpendicular axis. This property exists for the calamitics and the discotics, in the nematic phases, and the tilted smectic calamitics, such as the ferroelectric smectic-C*.

THE ORDER PARAMETER

The phases described above depend on the degree of order of the molecules. To describe the degree of order between the isotropic liquid and the liquid crystal phases, as well as between the different liquid crystal phases among themselves, the order parameter for liquid crystals is described as,

$$S = \left\langle \frac{3}{2}\cos^2\theta - \frac{1}{2} \right\rangle \tag{1.1}$$

The definition of an order parameter is the quantity that measures how much more ordered one phase is with respect to another, both mathematically and thermodynamically. This is achieved by it having a non-zero value in the more ordered phase and a zero value in the less ordered phase.

Let us consider first the calamitic liquid crystals. Assume the liquid crystals are rods, characterized by a unit vector **a**, which describes its direction in space. The nematic director **n** will be taken as the z-axis of the laboratory frame (Figure 1.6). Thus, **a** can be defined as,

$$a_x = \sin\theta\cos\varphi$$

$$a_y = \sin\theta\sin\varphi \tag{1.2}$$

$$a_z = \cos\theta$$

If the rods are given by a function $f(\theta,\varphi)$ that gives the behavior at a certain θ and φ, then one can define the state of alignment of

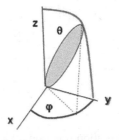

FIGURE 1.6 Liquid crystal rods oriented at an angle θ from the z-direction, taken as the director direction and at an angle φ from the x-y axes in the experimental direction.

the entire set of molecules by a distribution function over all the angles θ and φ. We find that this function varies from 1 at θ=0° to 0 at θ=90°. If we assume that the directions **a** and −**a** or **n** and −**n** are equivalent, then the function is again equal to 1 at θ=180°. This function is an even function and illustrated in Figure 1.7a. By examining the Legendre polynomials, we can see the similarity between the order parameter and the Legendre polynomial,

$$P_2(\cos\theta) = \frac{3}{2}\cos^2\theta - \frac{1}{2} \tag{1.3}$$

Since we are averaging over all the molecules in the sample, we recover equation 1.1.

The order parameter S relates the laboratory indices x, y, z to the molecular indices **a, b** and **c**. It is a traceless tensor, and symmetrical with respect to x, y, z and **a, b, c**. We have presented the order parameters as scalars but because the liquid crystals are anisotropic, they are tensorial quantities. Thus

$$S = \begin{pmatrix} S_{xx} & 0 & 0 \\ 0 & S_{yy} & 0 \\ 0 & 0 & S_{zz} \end{pmatrix} \tag{1.4}$$

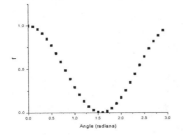
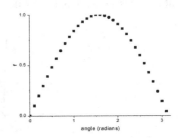

FIGURE 1.7 (a) The distribution function for a liquid crystal that satisfies **n** = −**n** and the value f≠0 for θ = π /2, 3π/2,... and f=0 for θ = π/2, 3π/2....; (b) the distribution function for discotic liquid crystals, which satisfy f = 0 for θ = 0, 2π, ... and f ≠ 0 for θ = π/2, 3π/2....

for the isotropic-nematic transition. This means that all off-axis components are zero for a calamitic nematic, $S_{xx}=S_{yy}$, and $S_{zz}=-2S_{xx}=-2S_{yy}$. We have not written the numbers for a, b, and c.

The distribution for a discotic liquid crystal is the opposite than for a calamitic liquid crystal, as shown in Figure 1.7b, with the director being perpendicular to the discotic molecule. Based on the distribution function shown, $f(\theta,\phi)S_{11}=-1/2$, $f(\theta,\phi)S_{22}=-1/2$, $f(\theta,\phi)S_{33}=0$. The order parameter for the ordered phase is given by S_{xx} or $S_{yy}=-1/2$.

The order parameter for the more ordered phases such as the smectics or the columnars is associated with the oscillation of the density of the center of mass of the molecules. This density oscillates for the smectics as the molecules move along the z-axis, such that,

$$\rho(z) = \rho_0 \left(1 + \psi e^{-ikz}\right) \tag{1.5}$$

where $|\psi|$ is the order parameter and the phase e^{-ikz} describes where the molecular layers are located along the z-axis. $k=2\pi/d$, and $\cos(2\pi z/d)$ is the real part of e^{-ikz}. $|\psi|$ measures the amplitude of the density oscillation. One can form a modified order parameter,

$$\left\langle \cos\frac{2\pi z}{d} \right\rangle \left\langle \left(\frac{3}{2}\cos^2\theta - \frac{1}{2}\right) \right\rangle \tag{1.6}$$

where $\cos\dfrac{2\pi z}{d}$ locates the layers. A better modified order parameter considers the density oscillations, ψ.

$$\left\langle |\psi|\cos\frac{2\pi z}{d} \right\rangle \left\langle \left(\frac{3}{2}\cos^2\theta - \frac{1}{2}\right) \right\rangle \tag{1.7}$$

Let us consider a transition between a nematic and smectic to get an idea of how to determine $|\psi|$.

Even though the nematic is not ordered in layers, the shape of the molecules and their orientation gives it a slight variation in density, resulting in a correlation between the liquid crystal molecules. This correlation can be modeled as a decaying exponential,

$$g(z) \propto \exp\left(-\frac{z}{\xi}\right) \tag{1.8}$$

where ξ is the correlation length of the molecule, that can be measured by X-ray or neutrons or light.

The Fourier transform for equation 1.8 is a Lorenztian:

$$g(z) \Rightarrow \frac{A}{\left[1+\left(k\xi\right)^2\right]} \tag{1.9}$$

$|\psi|$, which is proportional to equation 1.9 measures how many layers there are in a single "crystallite" on average at the molecular level, and ξ measures the way the molecules correlate to each other (for X-rays and neutrons), as shown in Figure 1.8. We note from equation 1.9 that $|\psi| = |\psi(\xi)|$. ξ can be subdivided into $\xi_{\parallel}$ and $\xi_{\perp}$ to determine the correlation of the molecules both parallel to the director and perpendicular to the director for well-aligned samples, consistent with the anisotropy of the liquid crystal. All three quantities will increase the closer the transition is to the smectic. Transitions into a tilted smectic will have tilt angle, ϕ as another order parameter.

The order parameter for the nematic–smectic or the smectic–smectic transition is not a scalar, as it is for the isotropic-nematic transition shown in equation 1.4, but because the liquid crystal is anisotropic, it is better described by a tensor. Thus, we have,

$$\xi = \begin{pmatrix} \xi_{xx} & 0 & 0 \\ 0 & \xi_{yy} & 0 \\ 0 & 0 & \xi_{zz} \end{pmatrix} \tag{1.10}$$

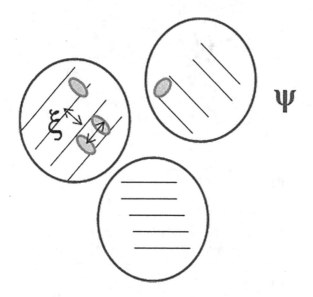

FIGURE 1.8   Geometrical meaning for the quantities ψ and ξ.

This assumes that the z-axis is aligned with the director for the liquid crystals. We shall see how these quantities are related to the energy of the liquid crystal in Chapter 2.

LIQUID CRYSTAL ELASTIC DEFORMATIONS

The undeformed ideal liquid crystal has the director pointing in the same direction, except in the chiral phases. The deformation is similar to distorting a spring, and Hooke's Law is applied,

$$E = \frac{1}{2}k(\Delta x)^2 \tag{1.11}$$

The distortions are on the order of micrometers, and thus can be detected with visible light. The deformation is related to the curvature of liquid crystals or the amount the liquid crystal deviates from being non-deformed. There are three main deformations, namely, splay, twist, and bend as shown in Figure 1.9 for the nematics. The energy due to the deformations is,

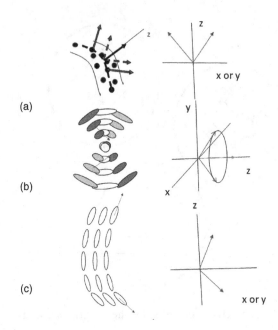

(a)

(b)

(c)

FIGURE 1.9 (a) The splay deformation for nematics. The dark gray "vectors" point in the directions where the molecular axis point which fall in the same plane; (b) the twist deformation for nematics. The dark gray "vectors" point in the directions where the molecular axis point, which fall in different directions that do not fall in the same plane; (c) the bend deformation for nematics. The dark gray "vectors" point in the direction where the molecular axis point which again fall in the same plane.

$$E = \frac{1}{2}K_1(\nabla \cdot \vec{n})^2 + \frac{1}{2}K_2(\vec{n} \cdot \nabla x \vec{n})^2 + \frac{1}{2}K_3(\vec{n}x\nabla x\vec{n})^2 \quad (1.12)$$

The constants K_1, K_2, and K_3, are the three main elastic constants that correspond to the splay, twist, and bend distortions. They are equivalent to the spring constant shown in equation 1.11. There are other distortions if we consider chirality or if we consider the proximity to a surface. If we have a chiral liquid crystal, there is a linear term to consider, $k_2(\vec{n} \cdot \nabla x \vec{n})$ and if we have a surface, there are terms such as the splay-saddle point K_{24}. We are considering the main distortions in the following discussion.

We now explain how the divergence terms that multiply these constants come about. Let us consider the spatial derivatives of **n**. They can be expressed as a tensor whose members are of the form,

$$\frac{\delta n_j}{\delta x_i}, \quad \text{where } i, j = x, y, z \qquad (1.13)$$

The tensor consists of a symmetric part,

$$\frac{1}{2}\left(\frac{\delta n_j}{\delta x_i} + \frac{\delta n_i}{\delta x_j}\right) \cong \nabla \cdot \vec{n} \qquad (1.14)$$

and an antisymmetric part,

$$\frac{\delta n_j}{\delta x_i} - \frac{\delta n_i}{\delta x_j} = \nabla x n_k \qquad (1.15)$$

If we consider that **n** is a unit vector and assume that it is in a frame of reference where z is parallel to n, all gradients of n_z vanish. We are not trying to change the length of n but to vary it from its original direction. We also note that,

$$(\nabla x \vec{n})^2 = (\vec{n} \cdot \nabla x \vec{n})^2 + (\vec{n} x \nabla x \vec{n})^2 \qquad (1.16)$$

The square of the antisymmetric part can be subdivided into a parallel component, $\vec{n} \cdot \nabla x \vec{n}$ and a perpendicular component, $\vec{n} x \nabla x \vec{n}$. Let us see how the molecules point out for the three distortions.

Let us consider the splay first. If n is along the z-axis, the variation along that axis is zero since the splay and indeed none of the distortions vary the length of *n*. Once the splay distortion is applied, the vectors along n_z points are shown in Figure 1.9a. These vectors have a projection along the *x–y* axis, without mixing n_x or n_y with z or any other axis. Therefore, this distortion can be expressed as,

$$\frac{\delta n_x}{\delta x} + \frac{\delta n_y}{\delta y} = \nabla \cdot \vec{n} \qquad (1.17)$$

We do the same with the twist distortion. To view what happens when the sample is twisted, we draw arrows to indicate where δn_x and δn_y are pointing and put them together as shown in Figure 1.9b. We note that the arrows move in a circle around the z-axis, where **n** points, emphasized by the central (light gray) arrow. Since the variations go around a circle, the divergence mixes the x- and the y-terms; therefore it is antisymmetrical. The twist is proportional to,

$$n_z\left(\frac{\delta n_y}{\delta x}-\frac{\delta n_x}{\delta y}\right)=\left(\vec{n}\cdot\nabla x\vec{n}\right)^2 \tag{1.18}$$

which corresponds to the parallel component in equation 1.16. Finally, we consider the bend distortion, as shown in Figure 1.9c. We do the same that we did with twist, by drawing arrows that show the direction where δn_x and δn_y points. We bring them together and notice that the resulting distortion is perpendicular to **n** and mixes the x- and y- terms as shown in Figure 1.9c. Comparing Figure 1.9c and Figure 1.9a, the variations in the bend are contained in a plane instead of a circle as shown for the twist (Figure 1.9b). Therefore, we again have an antisymmetrical term, but we consider the perpendicular component,

$$\left(\vec{n}x\nabla_{x}x\vec{n}\right)^2 \tag{1.19}$$

Collecting all three terms and multiplying them by the corresponding K_i's, one recovers equation 1.10. In general, K_3 is larger than K_1 or K_2, for a calamitic, because bending tends to force the molecules to come together in one side of the sample.

Let us examine the effect of each of the distortions on the smectics, again in the calamitics. When looking at the effects in smectics, we look at the effects on the layers. In Figure 1.10, we show the effects of the three distortions on the molecular layers. We see that under splay (Figure 1.10a) the layers remain unaltered, whereas under twist and bend, the layers are affected. Twist distorts the

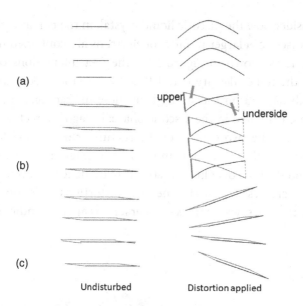

upper
underside

Undisturbed   Distortion applied

FIGURE 1.10   (a) Effects of the splay distortion on the calamitic smectic layers, leaving them unaltered. (b) Effects of the twist distortion on the calamitic smectic layers, having the same layer pointing in two different directions. (c) Effects of the bend distortion on the calamitic smectic layers. Bend brings the layers close to each other on one side and pulls them apart on the other side.

layers (Figure 1.10b). Notice the two molecules that have been drawn in the same layers. They point in different directions. Therefore, it can be said that the smectic phase does not support twist because it will tend to break it apart. Bend brings the molecules together as in the nematic (Figure 1.10c). It also distorts the layers, bringing them together, and therefore, it can be said that the smectic does not support bend. This has consequences in the phase transitions between the nematic and the smectic phases. All three values of the constants will increase near the transition, but the value of K_3 will increase at a larger pace. This can be detected through light scattering. The K's will play the role the ξ's play at the molecular level, and a relation between the K's and the ξ's can be derived.

Consider now the discotic liquid crystal, in the nematic phase. In the discotic, a larger number of electrons are contained in the x–y plane as shown in Figure 1.11. The π–π interactions of the face of the molecules are parallel to the director **n**. Any distortion will depend on the strength of these interactions, and the structure exhibited in the discotic phase: hexagonal, rectangular, or oblique. The two-dimensional structures run parallel to the director as well. Any change in the π–π interactions, the anisotropy, and the type of columnar structure influence how much the energy increases or varies for the three distortions. This compares to the calamitics, where the π–π interactions are perpendicular to

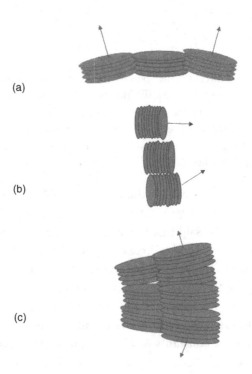

(a)

(b)

(c)

FIGURE 1.11 (a) Splay applied to a discotic phase. (b) Twist distortion for a discotic phase. (c) Bend distortion for a discotic. In all these figures, the dark gray arrows point in the directions that the discotic faces point.

the director. K_3 is not larger in general than K_2 or K_1 as was found for the calamitics.

If we apply splay to the nematic phase of the discotic phase, the two-dimensional separations in the structures that exist in the nematic will tend to run into each other, which will be energetically unfavorable (see Figure 1.11a). This movement of the two-dimensional structure will separate the molecules from each other. The strength of the π–π bonds will counteract this separation, increasing the energy. In some cases, the energy constant for the splay will saturate in the nematic phase. As the discotic phase is approached, both the two-dimensional structure and the bonding will counteract the splay distortion, and the energy will increase.

The twist distortion will have the faces of the molecules pointed in different directions about the central point (Figure 1.11b). The size of the molecules and the strength of the π–π interactions counteract this distortion. The strength of the π–π distortions will depend also on whether the liquid crystal is a lyotropic. As the molecules approach a more spherical shape, the distortion will not appear. This distortion will have a diverging behavior in the nematic and will increase as the discotic is approached.

Bend will push the short axes together and will distort the orientation of the columns (Figure 1.11c). In the small structures that exist in the nematic, the resulting energy is small. As the discotic phase is approached, the value for the energy will increase exponentially as the columns are assembled. The π–π interaction will have an increasing role as the discotic is approached.

The splay constant K_1 can initially be larger than the bend constant K_3 and at a point in the nematic when there are more and longer columns they can cross each other's values, since the splay brings together the faces of the disc together and the bend brings together the short axes. However, this is not true for all discotics. The twist constant K_2 is not easily comparable to the K_1 and K_3 because it depends on the π–π interactions and the anisotropy. The π–π interactions depend on the size and the shape of the discotics.

At high temperatures, when the columns in the nematic are not very large, the π–π interactions are not very large, and the value of K_2 may be below K_1 and K_3. It will increase as the columns grow, and as the discotic phase is approached, it will increase in value.

If we consider the descriptions for the ordered phases for both the nematic and ordered phases, we can predict that the bend and the twist for the calamitics, and the bend, the twist and the splay for the discotics tend to increase as a function of temperature as one approaches the ordered phases from the nematic. In this way, the behavior of the distortion constants can be considered as an order parameter.

DISCLINATIONS

In addition to the variations in the director, there are places where the director is not defined in the plane where we are observing the liquid crystal. These are defects, which can be point lines or sheets (also known as walls). The most common defects observed in liquid crystals are line defects, known as disclinations. The latter term was coined by Frank, and it originates from the Greek word, kline, which means slope, as illustrated on the top of Figure 1.12. A +1/2 disclination is used as an example here. The projection into the x–y plane is shown in the bottom part of the same figure.

The disclinations have the same mathematical form as the screw and edge dislocations, but instead of the shear modulus as the constant of proportionality, we consider the K's which we studied in the previous section, coupled to what is known as the "strength" m of the disclination, rather than the burger's vector, **b**. The energy associated with these dislocations is not defined at the point of the dislocation. In the vicinity of the screw dislocation, it is proportional to the natural logarithm of the distance from the dislocation. In the same manner that the dislocation energy for solids is given by,

$$E \propto \frac{Gb^2}{4\pi} \ln\left(\frac{R}{r_0}\right)$$

(1.20)

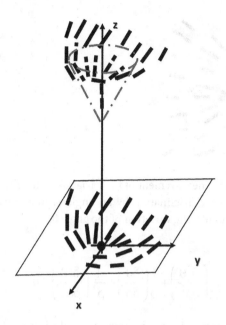

FIGURE 1.12   A +1/2 disclination in three dimensions and its projection into the x–y plane.

the free energy for a disclination is given by,

$$F_L = \pi K m^2 \ln\left(\frac{\rho_2}{\rho_1}\right) \qquad (1.21)$$

We consider the region where the K's all have the same value. This is true in the region close to the nematic–isotropic phase transition. The value for the free energy is minimized as one traverses a pathway described by the angle ϕ around the disclination point, as shown in Figure 1.13. The unit vector **n** at every value of ϕ varies as a function of the angle θ which depends on the position (x, y) of Cartesian coordinates or (ρ,ϕ) of cylindrical coordinates in the plane. When the free energy as a function of length F_L is minimized it satisfies Laplace's equation in two dimensions, which in cylindrical coordinates can be expressed as,

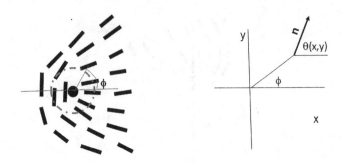

FIGURE 1.13 The measurement of the variation of the director **n** about the center of a +1/2 disclination by an angle ϕ, in a constant path of radius ρ. Right panel, the variation of the angle θ.

$$\left(\frac{\delta^2\theta}{\delta\rho^2}\right)+\frac{1}{\rho}\left(\frac{\delta\theta}{\delta\rho}\right)+\frac{1}{\rho^2}\left(\frac{\delta^2\theta}{\delta\phi^2}\right)=0 \tag{1.22}$$

Since the solution is found at fixed ρ, a solution to this equation that does not depend on ρ is,

$$\theta(\phi)=m\phi+\theta_o \tag{1.23}$$

where $\theta(\phi)$ changes by a multiple of π everytime ϕ changes by 2π. Therefore, $m=\pm\frac{1}{2},\pm1,\pm\frac{3}{2}...$, and is known as the strength of the disclination. If equation 1.23 is substituted in equation 1.22, then we obtain $\frac{1}{2}K\frac{m^2}{\rho^2}$, which can be integrated to obtain F_L,

$$F_L=\int_{\rho_1}^{\rho_2} 2\pi\rho d\rho\frac{1}{2}K\frac{m^2}{\rho^2}=\pi Km^2\ln\left(\frac{\rho_2}{\rho_1}\right) \tag{1.24}$$

The force between two disclinations can be attractive or repulsive depending on whether they are of equal value of m or of opposite value. This can be seen in Figure 1.14, where the force between

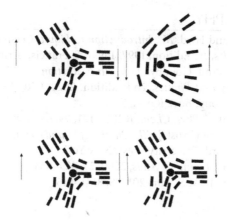

FIGURE 1.14 The forces between two disclinations of strength m=−1/2 and m=+1/2 and two disclinations of strength m=−1/2.

opposite disclinations ±1/2 or equal disclinations −1/2 are compared. The top part of Figure 1.14 looks at the interaction of a disclination of strength −1/2 and one of strength 1/2. The arrows show the direction of increasing φ for the two disclinations. The arrows add in between the two disclinations. The situation is the opposite for the dislocations that have the same strengths of −1/2. The arrows cancel each other between the two disclinations. The opposite happens when we examine the arrows outside the contact region. The arrows subtract for the upper panel of Figure 1.14 and add for the lower panel. They represent two forces of opposite sign. The upper panel represents an attractive force, while the bottom represents a repulsive force.

The interactions suggest that with an appropriate choice of substrate, the distribution and the distances of the disclinations can be controlled. The structure of nanomaterials mixed with the liquid crystals, and their corresponding properties can be controlled in this way. This may affect the behavior of hybrid photovoltaics, and organic photovoltaics at the boundary between the liquid crystal and the nanoparticles.

BIBLIOGRAPHY

Collings, P. J. and Hird, M., *Introduction to Liquid Crystals; Chemistry and Physics*, 1st Edition, 1997, Taylor & Francis, Bristol, PA.

Collings, P. J. and Goodby, J. W, *Introduction to Liquid Crystals: Chemistry and Physics*, 2nd Edition, 2020, CRC Press, Taylor & Francis Group, Boca Raton, FL.

Dastan, A., et al., *J. Phys. Chem. B* 2017, **121**, 9920–9928.

DeGennes, P. G. and Prost, J., *The Physics of Liquid Crystals*, 2nd Edition, 1993, Oxford University Press, New York, NY.

Hauser, A., et al., *Liquid Crystals*, 2000, **27**, 629–634.

Takacs, C. J., et al., *ACS Nano* 2014, 8, 8141–8151

CHAPTER 2

Energetic Properties of Liquid Crystals: Relation to the Order Parameters

W E CALCULATE THE ORDER parameters from the point of view of the energy of the liquid crystal in order to associate the energy of the liquid crystals with some of the properties we are interested in studying. We show the dependence of several liquid crystal properties. At a specific temperature, the properties can change as a function of applied fields, specifically, electric fields.

SELF-ASSEMBLY

Liquid crystals are one of many systems that self-assemble. Self-assembly occurs when the molecules that constitute them come together without any external manipulation, resulting in a larger supramolecular structure. The structure is stable but dynamic. It is subject to forces such as hydrogen bonds, π–π interactions, electromagnetic, and van der Waals interactions. When the configurations they adopt reach an equilibrium state, their free energy is said to be minimized. We can associate this minimized energy with the order parameters for the liquid crystals.

ENERGETIC CONSIDERATIONS

In Chapter 1, we introduced the different order parameters, S, ψ, ξ, and ϕ. All of these explain the structural changes that occur close to or at a phase change. We also introduced the constants K for the different deformations. The K's also vary as a function of how close the phase change is. The phase changes occur in general as a function of temperature, but they can also occur as a function of applied field or pressure or the change in concentration of a solvent. These changes can occur in an application, even when the temperature remains the same. In this section, we explore how the structural changes are explained in energetic terms, as a function of temperature.

One of the simplest theories to consider is the Landau-De Gennes theory, which was originally adapted and developed by de Gennes for the isotropic–nematic transition in liquid crystals, based on Landau's general description of phase transitions. Landau's description of a phase transition is based on minimizing the Helmholtz free energy,

$$F = E - TS \tag{2.1}$$

The Helmholtz free energy is related to the Gibbs free energy G through,

$$G = H - TS \tag{2.2}$$

where the enthalpy is given by $H = E + PV$. Therefore,

$$G = F - PV. \tag{2.3}$$

We are not considering the changes in phase as a function of pressure or of volume. Therefore, the change in the Gibbs free energy is the same as the change in the Helmholtz free energy. Consider first the phase change from the isotropic to the nematic. We then consider the change of the Gibbs free energy as a function of the order parameter S. To do this, we consider that close to the phase transition, S is very small, and the energy can be expressed as a sum of powers of S, where the linear term is not allowed because the energy of the lower temperature phase (in this case, the nematic) would be less than the energy of the higher temperature phase (the isotropic),

$$G(S,T) = G_{iso} + \frac{1}{2}A(T)S^2 + \frac{1}{3}BS^3 + \frac{1}{4}CS^4 + \cdots \qquad (2.4)$$

$C > 0$ to allow for the energy to have a minimum. Also, note that in considering a second-order transition for this case, the parameter $B = 0$. $G(S, T)$ is given by,

$$G(S,T) = G_{iso} + \frac{1}{2}A(T)S^2 + \frac{1}{4}CS^4 + \cdots \qquad (2.5)$$

For the isotropic–nematic transition, this occurs when we consider a two-dimensional nematic. $A(T) = 0$ at $T = T_c$ for both equation 2.4 and equation 2.5, at which point the free energy of the isotropic or high-temperature phase must be equal to the free energy of the nematic, or the low-temperature phase,

$$G(S,T)_{iso}\big|_{T_c} = G(S,T)_{nematic}\big|_{T_c} \qquad (2.6)$$

In addition, $G(S, T)$ is a minimum or,

$$\left(\frac{dG}{dT}\right)_s = 0 \qquad (2.7)$$

Consideration of equations 2.6 and 2.7 yields the values of the critical order parameter S_C at the transition. The results of considering equation 2.4 yield two minima, obtained by solving equations 2.6 and 2.7, and given by,

$$S_C = 0 \quad \text{or} \quad S_C = -\frac{2B}{3C} \text{ at } T = T_C, \tag{2.8}$$

where $B < 0$. If one considers equation 2.5 and combine it with equation 2.7, then,

$$S_C = 0 \quad \text{or} \quad S_C = \sqrt{\frac{A(T)}{C}} = 0 \text{ at } T = T_C \tag{2.9}$$

If we plot G as a function of S for the values given for S_C in equations 2.8 and 2.9, then we obtain Figure 2.1 for G as a function of S. At $T = T_C$, we see two minima for equation 2.8 and one minimum for equation 2.9, depicted in the cut lines. The local minima obtained from equation 2.8 depicted in the solid line have the same minimum. For a temperature $T < T_C$, the minimum corresponding to the nematic will become the lowest minimum. The temperatures in between will be temperatures where the isotropic and nematic will co-exist. At the point where the lowest minimum is the nematic one, we would find the second-order transition if there was one. The Landau-De Gennes theory and other related theories can be used to find out the relations between the order parameter and other quantities, such as $\Delta\varepsilon$, Δn, or σ, as we will see below.

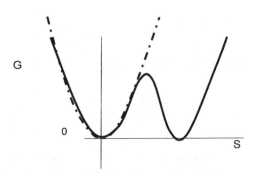

FIGURE 2.1 The minimum of the Gibbs free energy as a function of the order parameter for equation 2.8 (full line) and equation 2.9 (cut line).

THE SMECTIC PHASE

We have been considering the nematic phase as it is approached from the isotropic phase. We now consider the smectic phase as it is approached from the nematic phase. We investigate the signal generated by the smectic phase. The smectic phase is a layered phase along one direction in space, as shown in Figure 2.2. The number of layers is not as large as a solid crystal. The smectic phase is a crystal in one dimension with very few layers. It remains a liquid (fluid) for the other two directions. The X-ray (or neutron) signal associated with a crystal, given by $\sin^2 Nz/\sin^2 z$, where N is the number of layers, is approximated by a Gaussian,

$$\frac{\sin^2(Nz)}{\sin^2(z)} \rightarrow N^2 \exp\left(-\frac{(Nz)^2}{\pi}\right), \qquad (2.10)$$

when N is now a small number of layers. The Fourier transform of equation 2.10 is another Gaussian:

$$A\exp\left(-2\text{const}\left(k\xi\right)^2\right), \qquad (2.11)$$

where the correlation length ξ and the amplitude A are related to the number of layers N, and the relation $\Delta\xi\,\Delta k \cong 1$. We find that

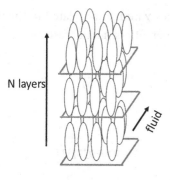

FIGURE 2.2 Sketch of the smectic A, showing the number of layers in one dimension and the fluid in the other two dimensions.

the amplitude for the Gaussian function maps to $|\psi^2|$ and ξ corresponds to the correlation length, ξ.

The Landau-De Gennes theory can be extended to the smectic phases, using the order parameter $|\psi|$ for the smectics. Similar to the order parameter for the nematics, close to the transition the value of the smectic order parameter is small. The energy that describes the formation of the layers coming from the nematic or the isotropic to the smectic phase can be expressed as,

$$G(\psi,T) = G_{nem/iso} + \frac{1}{2}\alpha(T)|\psi|^2 + \frac{1}{4}\beta|\psi|^4 + \frac{1}{6}\gamma|\psi|^6. \quad (2.12)$$

The order parameter for the smectic or discotic phases, $|\psi|$, can be related to the order parameter S by considering that as the smectic (or discotic) is approached, the value for the parameter S varies depending on the width of the nematic phase. Let δS be the distance to the smectic phase, which is included in equation 2.12 as,

$$G(\psi,T) = G_{nem/iso} + \frac{1}{2}\alpha(T)|\psi|^2 + \frac{1}{4}\beta|\psi|^4$$

$$+ \frac{1}{6}\gamma|\psi|^6 + \frac{1}{2}\frac{(\delta S)^2}{\chi} - \mu(\delta S)|\psi|^2 \quad (2.13)$$

where the quantities χ and μ are related to the width of the nematic phase. At the nematic-smectic transition temperature,

$$\frac{dG}{d(\delta S)} = 0. \quad (2.14)$$

or,

$$\delta S = \chi\mu|\psi|^2 \quad (2.15)$$

The value for δS will combine with the value for $\beta|\psi|^4$ to determine whether the transition is first or second order.

The density wave $|\psi|$ varies in space as the phase transition is approached. The density wave can be expressed in terms of a Fourier component:

$$\psi = \int e^{i\vec{q}\cdot\vec{r}}\rho(q)\frac{d\vec{q}}{(2\pi)^3} \qquad (2.16)$$

The density wave will vary in $\mathbf{q}$ as the layers vary. Mathematically, this means that we must consider terms in $|d\psi/dr|^2$ in equation 2.12. We subdivide $\mathbf{q}$ into $\mathbf{q}_{\parallel}$ and $\mathbf{q}_{\perp}$ to consider the anisotropy of the liquid crystal, $\mathbf{q}_{\parallel}$ for the direction parallel along the long axis of the molecule and $\mathbf{q}_{\perp}$ for the direction along the short axis of the molecule. This means that the coefficient of the term in $d\psi/dr$ must be subdivided into $D_{\parallel}$ and $D_{\perp}$. If we keep the term in $|\psi|^2$ in equation 2.12 and add the term in $|d\psi/dr|^2$, ignoring the other terms for the time being, since $|\psi|^2$ is very small and close to the transition, we obtain,

$$G - G_{\text{nem}} = \frac{1}{2}\alpha(T)|\psi|^2 + \frac{1}{2}\left|\frac{d\psi}{dr}\right|^2$$

$$= \frac{1}{2}\alpha(T)|\psi|^2 + \frac{1}{2}\left(D_{\parallel}|q_{\parallel}|^2 + D_{\perp}|q_{\perp}|^2\right)|\psi|^2 \qquad (2.17)$$

The free energy must be partitioned equally between the oscillations. Thus,

$$\left\langle \frac{1}{2}\left(\alpha(T) + D_{\parallel}q_{\parallel}^2 + D_{\perp}q_{\perp}^2\right)|\psi|^2\right\rangle = \frac{1}{2}kT, \qquad (2.18)$$

which gives for $\left\langle |\psi|^2\right\rangle$,

$$\left\langle |\psi|^2\right\rangle = \frac{kT}{\left[\alpha(T) + \left(D_{\parallel}|q_{\parallel}|^2 + D_{\perp}|q_{\perp}|^2\right)\right]}$$

$$= \frac{kT}{\alpha(T)}\frac{1}{1 + \frac{1}{\alpha(T)}\left(D_{\parallel}|q_{\parallel}|^2 + D_{\perp}|q_{\perp}|^2\right)} \qquad (2.19)$$

The answer has the form of a Lorentzian, as we observed by assuming that the nematic had a decaying exponential in Chapter 1. The coefficients $D_{\parallel}$ and $D_{\perp}$ are related to the correlation lengths along the parallel and perpendicular directions,

$$\xi_{\parallel}^2 = \frac{D_{\parallel}}{\alpha(T)}; \quad \xi_{\perp}^2 = \frac{D_{\perp}}{\alpha(T)} \tag{2.20}$$

The way these order parameters vary as a function of temperature is more complicated than this theory predicts depending on the transitions studied and the liquid crystal phases involved. However, the derivation serves to illustrate how the order parameters depend on the temperature, even when the dependence is not given by a square root dependence. The parameters may mix the parallel and the perpendicular directions, depending on how the liquid crystal is ordered, and experimentally speaking, if the liquid crystal is in a polycrystalline state. We can compare how the order increases as a function of temperature as shown in Figure 2.3, which shows how a liquid crystal becomes more ordered as the nematic smectic. In addition, Figure 2.3 illustrates the effect of the proximity to a substrate has on the value of the order parameter, increasing the order at high temperatures, especially at temperatures close to the isotropic-nematic phase transition.

We can consider the oscillation in the long wavelengths (experimentally, the visible wavelengths) by considering an oscillation in the elastic constants, the same way that $|\psi|$ measures the changes at the molecular level. The energy can be written as,

$$G = \alpha(T)|\vec{n}|^2 + \frac{1}{2}K_1(\nabla \cdot \vec{n})^2 + \frac{1}{2}K_2(\vec{n} \cdot \nabla x\vec{n})^2 + \frac{1}{2}K_3(\vec{n}x\nabla x\vec{n})^2 \tag{2.21}$$

Considering the Fourier transform of the director,

$$n(q) = \int n e^{i\vec{q}\cdot\vec{r}} d\vec{r} \tag{2.22}$$

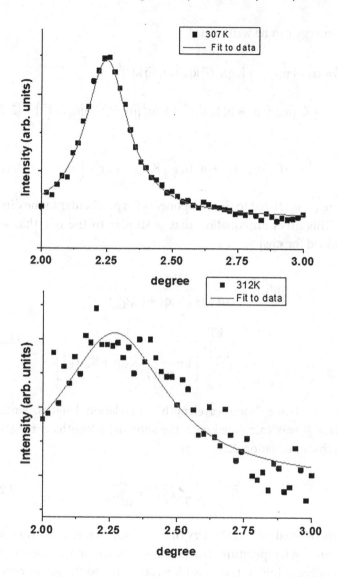

FIGURE 2.3 The variation of 8CB as a function of temperature close to the nematic–smectic transition temperature (left panel) and close to the nematic–isotropic transition temperature. The peak that is observed very close to the nematic–isotropic transition shows the effect of the proximity to the substrate.

The energy can be written as,

$$G = \alpha(T)\left|n(\bar{q})\right|^2 + K_1\left|n_x(\bar{q})q_x + n_y(\bar{q})q_y\right|^2$$

$$+ K_2\left|n_x(\bar{q})q_y - n_y(\bar{q})q_x\right|^2 + K_3 q_z^2\left(\left|n_x(\bar{q})\right|^2 + \left|n_y(\bar{q})\right|^2\right) \quad (2.23)$$

$$\alpha(T)\left|n_\alpha(q)\right|^2 + \left|n_\alpha(q)\right|^2\left[K_3 q_\parallel^2 + K_\alpha q_\perp^2\right] = kT \quad (2.24)$$

where K_α is related to the components perpendicular to the director. This gives an equation that is similar to the one that was obtained for $<|\psi|^2>$,

$$\left\langle\left|n(q)\right|^2\right\rangle = \frac{kT}{\alpha(T) + \left(K_3 q_\parallel^2 + K_\alpha q_\perp^2\right)}$$

$$= \frac{kT}{\alpha(T)}\frac{1}{\left[1 + \dfrac{1}{\alpha(T)}\left(K_3 q_\parallel^2 + K_\alpha q_\perp^2\right)\right]} \quad (2.25)$$

We can define "long-wavelength" correlation lengths, similar to how it was calculated with the short-wavelengths correlation lengths, in equation 2.20,

$$\xi_{K_3}^2 = \frac{K_3}{\alpha(T)};\ \xi_{K_\alpha}^2 = \frac{K_a}{\alpha(T)} \quad (2.26)$$

As mentioned before, the way that the order parameters vary as a function of temperature is more complicated than this theory predicts depending on the transitions studied and the liquid crystal phases studied, the same as the order parameters ψ and ξ.

LIQUID CRYSTAL PROPERTIES

The properties of the liquid crystal are related to the different order parameters presented above in their dependence on temperature.

The order parameter ξ the correlation length for a well-ordered liquid crystal is expressed as,

$$\overset{\leftrightarrow}{\xi} = \begin{pmatrix} \xi_\perp & 0 & 0 \\ 0 & \xi_\perp & 0 \\ 0 & 0 & \xi_\parallel \end{pmatrix}, \tag{2.27}$$

similar to the order parameter given for the nematics, S. The properties we are particularly interested in are the electronic and optical properties. If we consider the dielectric constant, ε it is expressed as,

$$\overset{\leftrightarrow}{\varepsilon} = \overline{\varepsilon} + \Delta\varepsilon, \tag{2.28}$$

where $\overline{\varepsilon}$ is the average dielectric constant and $\Delta\varepsilon$ is the anisotropy of the dielectric constant.

As the temperature decreases, the value of the measured $\varepsilon_\parallel$ is concentrated along the long axis of the molecule for calamitics. The measured $\Delta\varepsilon$ will depend on the number of molecules that align with the director on average. In its simplest form,

$$\Delta\varepsilon \sim S, \tag{2.29}$$

or more generally, $\Delta\varepsilon \sim S^i$. This is illustrated in Figure 2.4. The value obtained for $\Delta\varepsilon$ will depend on the particular molecule and the model used, but it always will have a dependence on the order parameter. In the smectics, it will depend on the order parameter ξ or $|\psi^2|$, at least for the smectic-A's. ψ is a measure of how many layers there are in a single "crystallite" as seen in Chapter 1. ξ corresponds to the correlation between molecules inside the crystallite. Therefore, if we add additional layers to the crystalline, the order parameters ξ and $|\psi^2|$ will increase and the value for $\Delta\varepsilon \sim \xi$ or in general, $\Delta\varepsilon \sim \xi^i$.

Similarly, the electrical conductivity can be expressed as a tensor. The electrical conductivity from the liquid crystals arises

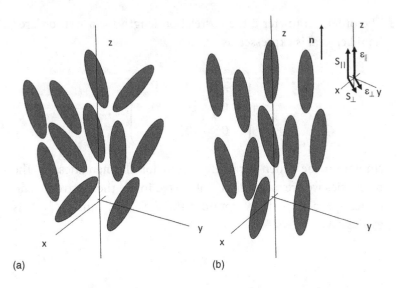

FIGURE 2.4 The dielectric constant illustrated in the isotropic (a) and the nematic (b). The dielectric constant ε mixes $\varepsilon_{\parallel}$ and $\varepsilon_{\perp}$ in the isotropic, whereas $\varepsilon_{\parallel}$ aligns along $S_{\parallel}$ while $\varepsilon_{\perp}$ aligns along $S_{\perp}$ as shown in the inset to the figure.

from the hopping of electrons along the delocalized electrons in the aromatic ring structures generally found in the central core of most liquid crystals. The aromatic rings face in a direction perpendicular to the director for the calamitics and are parallel to the director for the discotics. The conductivity will depend on how well the molecule is aligned as shown in Figure 2.5 for the calamitics and Figure 2.6 for the discotics. As the temperature drops, more molecules will line up in the direction of the director. In the nematic, this means that the electrical conductivity is related to the order parameter S. In the smectic, it is related to the order parameter $\xi_{\perp}$ for the calamitics and $\xi_{\parallel}$ for the discotics (equations 2.20 and 2.25). The two correlation lengths are related to $<|\psi|^2>$ (equation 2.19), which measures how large the "crystallite" is. This determines how far a carrier (electron or hole) can traverse without encountering a grain boundary or a defect that will scatter this carrier. A more ordered liquid crystal will

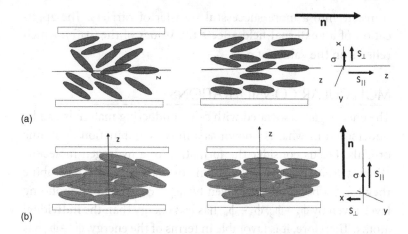

FIGURE 2.5 (a) The conductivity for a nematic calamitic liquid crystal. The sample is shown between two electrodes. The conductivity aligns with $S_\perp$. It aligns with $\xi_\perp$ for the smectics; (b) the conductivity for a nematic discotic liquid crystal. The sample is shown between two electrodes. The conductivity aligns with $S_\parallel$. It aligns with $\xi_\parallel$ for the discotic phases.

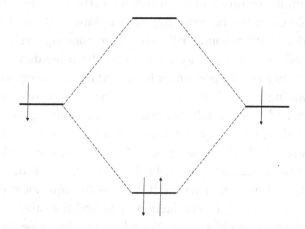

FIGURE 2.6 The formation of the energy gap arising from two hydrogen atoms coming together.

translate into a more successful transfer of carriers. The application of an external field affects the values of the $\langle |\psi|^2 \rangle$, which reflects on the values of σ.

MOLECULAR CONSIDERATIONS

The energy gap associated with semiconducting materials can be understood in what is known as a linear combination of atomic orbitals. Let us consider the formation of a hydrogen molecule. When two hydrogens, with atomic orbitals ϕ_A and ϕ_B, combine they form a bonding level given by $\phi_A + \phi_B$ and an antibonding level, given by $\phi_A - \phi_B$. $\phi_A + \phi_B$, has less energy than the individual atoms. Therefore, it is favorable in terms of the energy. $\phi_A - \phi_B$ has more energy than the individual atoms, and therefore it is unfavorable. Unfavorable means that the molecule will prefer to exist as two independent atoms. The formation of these two levels is shown in Figure 2.6. The energy between the two configurations constitutes a gap, where the level of the bonding energy is designated as the highest occupied molecular level (HOMO), and the level of the antibonding level is designated as the lowest unoccupied molecular state (LUMO) in this case.

A similar picture can be drawn for carbon in the benzene ring. Consider the benzene ring. The unsaturated carbon that gives rise to the π-bonds will have all sp^2 bonding levels at the lower values of the energy followed by the p-bonds that will combine to form the bonding π-bonds. At higher energy are the antibonding π^*-bonds that are empty. The benzene ring has six carbons and six p orbitals that result in three bonding and three anti-bonding levels as estimated from a Frost circle, as shown in Figure 2.7. The difference in energy between these two levels, the lowest unoccupied molecular level (LUMO) and the highest occupied molecular level (HOMO), is the equivalent of the difference between the conduction band and the valance band encountered in solid-state semiconductors. The value for this energy for the benzene ring is 2.4 eV.

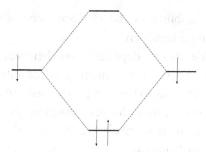

FIGURE 2.7 The formation of the HOMO and LUMO levels from the benzene ring as determined with a Frost diagram.

The mobility of the carriers in the molecules depends on their dipole moment and the molecular packing. The conductivity is related to the mobility of electrons and holes,

$$\sigma = e\left(n\mu_e + p\mu_p\right) \tag{2.30}$$

The conductivity is affected by the presence of ionic impurities that dominate in the more disordered phases of the liquid crystals. If we take the examples shown in Figures 2.6 and 2.7, the conductivity for electrons will be determined by the shape of the LUMO and for the holes will be determined by the shape of the HOMO.

Several aspects of the liquid crystal structure have to be considered to evaluate the electrical conductivity in these materials. Some of these are as follows: the alignment of the central aromatic cores; the substituents in the central core; the size and the length of the alkyl side chains; the donor or acceptor nature of the side chains; and the viscosity of the liquid crystal phase. We subdivide the following discussion between the discotics and the calamitics.

Discotics: The liquid crystal aligns about the central core. If it consists of an aromatic ring core, it provides the π–bonds that allow the electrons to hop from one molecule to the next. This depends on how close the aromatic cores are. The size of the core influences the mobility of the electrons since they provide a larger π-stacking interaction.

The alkyl side chains, depending on their size, increase the disorder of the system. The branching of the side chain or the addition of substituents to the side chain results in a reduction of the π-interactions due to the lower interaction between the cores. The side chains provide an insulating character that separates the columns of cores from each other and reduces the probability that the carriers may tunnel from one column to the next. These effects result in a one-dimensional carrier transport. The electro-negativity of the substituent may impede the carriers from tunneling. It may also determine the nature of discotic device, either being a donor or being an acceptor.

The lower the viscosity of the liquid crystal chain, the better the possibility to align the liquid crystal phase because it reduces the defects and the grain boundaries that impede the carrier transport. This is connected with the structural properties that were presented at the beginning of this chapter, such as $|\psi|$ and $\xi_\perp$

Calamitics: The calamitics have their cores oriented perpendicular to the director. Therefore, they have to be oriented with the director parallel to the substrate that holds them (Figure 2.5a). This orientation is similar to the one used in display devices. Because the cores are perpendicular to the director they are oriented along the directions where the liquid crystal exhibits liquid-like behavior, which suggests that a lower temperature smectic would be ideal to first order. This property of being associated with the liquid-like behavior separates the cores and makes the π–π interactions weaker. The conductivity depends on $\xi_\perp$ as mentioned before. Even in the nematic, where it is smaller than the smectics, the value of this parameter $\xi_\perp$ is larger than the value for the liquid. The cores can be made longer, which tends to stabilize the liquid crystal phase. A longer core increases the charge transfer. The proximity of the cores makes the carrier transport two dimensional. The two-dimensional aspect of the transport makes it more tolerant to defects as opposed to the one-dimensional transport obtained for the discotics. This increases the charge transport even when we consider the limitations cited above. Substituting a bulkier substituent in the cores can destabilize and even eliminate the liquid crystal phase.

The longer the alkyl side chains, the more likely that a smectic phase will be obtained. Longer alkyl chains on both sides of the molecule are preferred rather than a long alkyl chain on the one side and a considerably shorter on the other side. The alkyl chains, in general, have no substituents for this application. Substituents affect the stability of the smectic phase. The position with respect to the core and the size of the substituent can reduce or eliminate the smectic phase. The reduction in lamellar packing due to the size of the substituent can be counterbalanced with the polarity of the substituent.

The lower viscosity obtained in the nematic phase allows for the molecules to align. The viscosity helps in the self-healing of defects both in the smectic and in the nematic phases. However, the conductivity in the low viscosity nematic phase is dominated by the ionic impurities. This effect is reduced with decreasing temperature, and once the nematic transitions to the higher viscosity smectic, the electronic processes dominate and produce constant carrier mobilities.

Polymers: The alignment and corresponding conductivity and mobility produced by the discotics and the calamitics suggest that including them in polymers can provide the structural stability in addition to providing mechanical stability, which the liquid crystals lack as small molecules. However, the results for the conductivity can be mixed ones. This is due in part to the fact that the crystalline structure has defects and grain boundaries that interfere with the mobility and the conductivity, and to the decreased electron mobility compared to the single-molecule liquid crystal. To form liquid crystal polymers, a polymerizable group must be added to the alkyl side chain, which can affect the stability of the smectic phase, and hence of the resulting polymer. The stability of the smectic phase of the liquid crystal can be affected such that it disorders into the nematic.

If we consider a polymer with a side-chain liquid crystal, the length of the spacer between the polymer and the liquid crystal unit will determine the order of the liquid crystal and will modify the carrier mobility in the liquid crystal unit.

As mentioned before, we need to look at how the samples are prepared in order to obtain the best alignment of the liquid crystals. This will lead to a sample that has a minimum of grains ($<|\psi|^2>|$ maximized) and for the molecules that stack together such that $\xi_\perp$ is maximized for the calamitics (Figures 2.4 and 2.5a), $\xi_\parallel$ maximized for the discotics (Figure 2.5b).

REFERENCES

Branch, J., et al., *J. Appl. Phys.* 2004, **115**, 164313.

Collings, P. J. and Hird, M., *Introduction to Liquid Crystals; Chemistry and Physics*, 1st Edition, 1997, Taylor & Francis, Bristol, PA.

Collings, P. J. and Goodby, J. W., *Introduction to Liquid Crystals: Chemistry and Physics*, 2nd Edition, 2020, CRC Press, Taylor and Francis Group, Boca Raton, FL.

DeGennes, P. G. and Prost, J., *The Physics of Liquid Crystals*, 2nd Edition, 1993, Oxford University Press, New York, NY.

Hirst, L., *Fundamentals of Soft Matter Science*, 1st Edition, 2012, CRC Press, Boca Raton, FL.

Martínez-Miranda, L. J. and Hu, Y., *J. Appl. Phys.* 2006, **99**, 113522.

Martínez-Miranda, L. J., et al., *Appl. Phys. Lett.* 2010, **97**, 223301.

Matthews, J. and Walker, R. L., *Mathematical Methods of Physics*, 2nd Edition, 1970, W. A. Benjamin, New York.

Warren, B. E. *X-ray Diffraction*, 1969, Addison-Wesley Publishing Co., Reading, MA.

CHAPTER 3

Review of Organic Electronics

THERE ARE DIFFERENT MEANS of electron transport available for organic systems. In this chapter, we review band transport, hopping, and exciton dissociation, and how they are related to transport in organic compounds.

BAND TRANSPORT

The origin of the energy gap was introduced in Chapter 2 by comparing the electron levels for two atoms or for a molecule. The electron levels have different values due to the Pauli exclusion principle. In crystalline solid materials, the electrons are arranged in energy bands, as electrons adopt different energy values (see Figure 3.1). N is the number of crystalline boxes contained in the material and a is the one-dimensional crystal unit cell value. The conduction electrons will interact with the ion cores of the crystal. The potential energy of the ions attracts and slows downs the electrons that lead to the appearance of energy gaps or band gaps, as shown in Figure 3.2. This interaction of the conduction electrons with the ion cores will follow the periodic lattice of the solid.

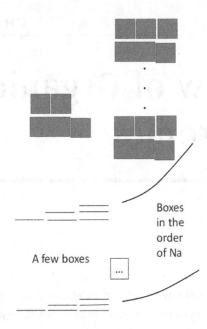

A few boxes

Boxes in the order of Na

FIGURE 3.1   Sketch illustrating the origin of the energy bands in solids.

There are no electron levels available in the energy gaps, at least in the pure samples. The gaps arise near the Fermi energy, the highest energy level occupied by electrons in the material. Similarly, if we consider a system consisting of many molecules, the different electron levels will expand into bands of energy.

Charge transport depends on the type of material. We concentrate on the semiconductors. Semiconductors are similar to the insulators, in that at 0°K, all available electron energies are occupied up to the energy gap. However, for semiconductors at T≠0°K, there is a possibility that electrons may be excited to the higher states in the conduction band, because the energy gap is typically in the order of 1eV. The excited electrons will be created at a rate proportional to $\exp(-\Delta E/k_B T)$ at any temperature T≠0°K. This implies there will be an equilibrium concentration of carriers that conducts electricity. The excited electron is surrounded by empty states where an electric field can draw the electron into a conduction process.

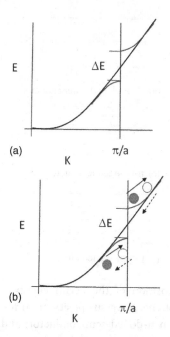

FIGURE 3.2 (a) The slowing down due to the ions in the cores that give rise to the energy gaps is shown in dark gray. (b) The movement of electrons (straight arrows, dark circles) and holes (cut arrows, light circles) in both the conductive and valence band.

The excited electron leaves a vacant state in the valence band. This vacant state is essentially the absence of an electron, known as a hole. The neighboring electrons can move into this vacant state due to the same electric field, and the electrons both in the conduction and in the valence band will move in the same direction within the band due to the electric field. However, the vacant states or holes move in the opposite direction, carrying with it a net positive charge, shown in the cut-line arrow in Figure 3.2b. Because at any non-zero temperature, there are excited electrons in the conduction band, the Fermi level for pure, non-doped semiconductors is placed in the middle of the energy gap. However, the number density of these intrinsic carriers is low (Figure 3.3a).

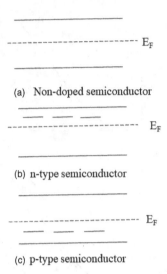

FIGURE 3.3 The position of the Fermi energy for semiconductors shown in a graph of energy versus position: (a) for an intrinsic semiconductor; (b) for an n-doped semiconductor; and (c) for a p-doped semiconductor.

If impurities are added to the intrinsic semiconductor in such a way that they substitute for the semiconductor atom in the crystal structure, new electron levels are created in the gap since all electron levels were completely occupied by the electrons of the intrinsic semiconductor. If the impurity has one electron in excess of the semiconductor structure, the energy level for that excess electron will be slightly lower than the conduction band, and located in the energy gap, where there were no electronic states (Figure 3.3b). This constitutes an n-type (donor impurity) semiconductor. If the impurity has one electron less than the semiconductor structure, the extra hole will occupy an electron level which is slightly higher than the valance band, located at the energy gap. This constitutes a p-type semiconductor (Figure 3.3c).

The energy difference between these levels and the bands will be much smaller than the band difference ΔE. Therefore, it will be easier to excite an electron or a hole into the corresponding band

and make them available for conduction. Because of the extra energy levels, the Fermi level will move from being at the center between the valence and the conduction levels, to being displaced more toward the conduction band for the n-type semiconductors (Figure 3.3b) and more toward the valance band for the p-type semiconductors (Figure 3.3c). The Fermi level has to have the same value when the n- and the p-type semiconductor appear in the same system. This causes a bending of the bands if we have a p-n junction, as shown in Figure 3.4.

The levels close to the Fermi Energy are the equivalent of the HOMO and LUMO levels in the organics (see Figures 2.6 and 2.7). Just as in the solid-state semiconductors, there will be an equilibrium amount of electrons and holes available for conduction. Usually, holes just under the HOMO level are mostly responsible for the transport in organics, since electrons above the LUMO

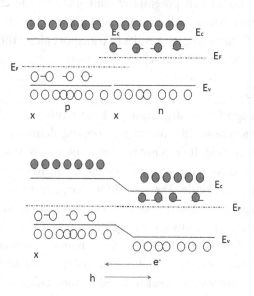

FIGURE 3.4 An n-p junction, in sketch illustrating how the Fermi level aligns itself. Upper panel: just before they are put together; lower panel: when the n and p-type semiconductor join. The Fermi energy is the same for the entire system and causes a bending of the bands.

FIGURE 3.5 Sketch showing the intrachain motion of the electrons (straight arrow).

level are stopped by traps. The traps are characterized by an activation energy ε_A, and the mobility is expressed as,

$$\mu = \mu_o \exp-\left(\frac{\varepsilon_A}{k_B T}\right), \tag{3.1}$$

where ε_A can be seen as the depth of the trap. Traps may be defects, or ions or related to the way the compounds terminate at surfaces. Polarons that have energies around the $k_B T$ above the HOMO or below the LUMO can potentially contribute to the charge carrier transport. These will proceed along the chain as shown in Figure 3.5. Band transport results in transport along the chains of the material, that is, in intrachain transport.

HOPPING

The most important charge transport mechanism for organics, be them polymers or small molecules is hopping, defined as thermally activated tunneling. It does not require a conjugated polymer since carriers can move from one part of the chain to another due to defects in the molecule. The model for charge transport based on band transport described above is based on inorganic semiconductors, where the atoms are strongly localized. We review several theories in this section that describe this phenomenon for polymers and small molecules, such as non-polymeric liquid crystals.

Polymers consist of weakly bonded molecules with a large degree of conformational freedom. The arrangement of these molecules varies from being amorphous to semicrystalline. The limit for charge transport is the connection and relation between the semicrystallite regions, with a high degree of π–π stacking dispersed in a disordered matrix. A typical crystalline packing involves lamellae of co-facially stacked molecules that exhibit π–π overlap. The disorder is related to the existence of electronic traps that limit charge transport. Some polymers and small molecules have hopping occur through the π–π bonds of the aromatic rings or the π–double bonds that constitute part of their structure.

Because of the variation in energies and distances in many organic semiconductors, a Gaussian distribution density of states is assumed,

$$g(\varepsilon) = \frac{N}{\left(2\pi\sigma_g\right)^2} \exp\left(-\frac{\varepsilon^2}{2\sigma_g^2}\right), \qquad (3.2)$$

where σ_g is the width of this density of states. The filled states available for charge transfer are those that are located at the top of this Gaussian variation. The mobility μ depends on this Gaussian distribution and on the electric field according to the Poole-Frenkel equation,

$$\ln\mu \propto E^{\frac{1}{2}}, \qquad (3.3)$$

where E is the applied electric field, which can be included in the Gaussian, as shown in equation 3.4, where the coefficient of $E^{1/2}$ is the inverse of the width of the position. An equation for μ that includes the Gaussian variations in energy (equation 3.2) and the variations in position Σ_o (equation 3.3) is,

$$\mu = \mu_o \exp\left(-\frac{2}{3}\left(\frac{\sigma_g}{k_B T}\right)^2\right) \exp\left(\left(\left(\frac{\sigma_g}{k_B T}\right)^2 - \Sigma^2\right)\sqrt{E}\right). \qquad (3.4)$$

Equation 3.4 is related to the conductivity σ for disordered semiconducting materials, by the experimental relation,

$$\sigma \propto \mu^\delta \qquad (3.5)$$

or

$$\sigma = ne\mu^{\delta}. \tag{3.6}$$

The mobility μ depends on the rate of transfer as expressed by Marcus and a modified Miller-Abrahams formalism as,

$$\omega_0 \left| H_{ij} \right|^2 \exp\left(-\frac{\Delta G - \lambda}{kT} \right), \tag{3.7}$$

where ΔG is the Gibbs free energy, given by the difference between the valence band and the conduction band or the HOMO and LUMO levels of the polymer, and H_{ij} is the transfer integral between the states i and j. λ is the reorganization energy for the nuclei. The value of λ depends on the values for the ground state and the first excited states of the nuclei. It involves two nuclei: the donor molecule that changes from a charged state to a neutral state and the acceptor molecule that changes from a neutral to a charged state. The mobility μ can be expressed in terms of the hopping rate between level i and level j, ω_{ij}, the probability that these levels be occupied, p_i, and the distance R_{ij} between the levels along the direction of the applied field, and the total number of available levels

$$\mu = \frac{\sum\limits_{i,j,i\neq j} \omega_{ij} p_i \left(1 - p_j\right) R_{ij,E}}{pN_i E}. \tag{3.8}$$

The efficiency depends on how the π–π bonds are stacked or misaligned, and how well these structures form percolating networks. Short polymers will have quasiperfect crystals separated by a disordered region. The percolation is limited by the proximity of the lamellae to each other or the size of the disordered region that separates the quasicrystalline regions (see Figure 3.6, upper panel). As the length of the polymer grows, the semicrystalline

FIGURE 3.6 The percolation in both a short polymer (upper panel) and a longer polymer (lower panel). The percolation between very close π–π bonds (straight lines), and between bonds that are farther away (cut lines). (Based on Rolland, Physical Review Materials 2, 045605 (2018).)

regions are not so well localized as shown in the upper panel of Figure 3.6, and the percolation will depend on how close the π–π bonds lie (Figure 3.6, lower panel). The percolation increases due to the increased transfer efficiency of the π–π bonds. As the length of the polymer increases the percolation saturates.

Another way of determining the mobility and conductivity is the diffusion constant. The arrangement of the quasicrystal-line region can be compared with the polymer adopting nematic order in conjugated polymers. The charge carriers are governed by the π–π interactions as mentioned before, and how close they are positioned with respect to each other. If $\left(\left\langle R^2 \right\rangle\right)^{1/2}$ is a measure of this average distance between the π–π bondings between two aromatic rings or the expression of the degree of overlap of their wave functions, we can express the diffusion constant as,

$$D = \lim_{\tau \to \infty} \frac{\left\langle R^2 \right\rangle}{2\tau}, \qquad (3.9)$$

where τ is the characteristic time for a charge, be it an electron or a hole, to diffusively spread either along a chain or across successive chains. The charge will diffuse without slowing down until it finds a defect that acts as a trap. In three dimensions, we can consider the size of a grain as this trap. If the charge transfer is considered a one-dimensional random walk of step length $\left(\langle R^2 \rangle\right)^{1/2}$, the mobility according to the Einstein relation is,

$$\mu = \frac{e\langle R^2 \rangle}{2k_B T\tau}. \tag{3.10}$$

The arrangement of the quasicrystalline region inside the polymer can be compared to the polymer adopting nematic order (Figure 3.7). This means that the mobility is anisotropic when the direction along which the chains are extended is compared to the direction perpendicular to it. The perpendicular direction is the direction perpendicular to the lamellar structure as seen in Figure 3.7 and is associated with the hopping direction. $\langle R^2 \rangle$ is expressed in terms of the nematic order parameter S when polymers are submitted to uniaxial alignment, given by the order parameter, presented in Chapter 1, and rewritten here,

$$S = \frac{3}{2}\langle \cos^2 \theta \rangle - \frac{1}{2}. \tag{3.11}$$

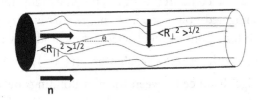

FIGURE 3.7 A polymer close to a surface exhibits nematic order along the perpendicular average distance separating the polymer chains. The angle θ measures the deviation from the nematic director, which is taken along $\langle R_\perp^2 \rangle^{1/2}$.

The perpendicular direction depends on the fluctuations, $\Delta r_\perp$, constrained by the nematic field. The $\Delta r_\perp$'s depend on a deflection length λ and are inversely proportional to the persistence length l_p of the polymer,

$$\Delta r_\perp \approx \frac{2\lambda^2}{l_p}. \tag{3.12}$$

The fluctuations depend on the stiffness of the chains below a certain value of λ. If l_p is larger than λ, the nematic order can be approximated by,

$$S \approx 1 - \frac{3}{2}\left(\frac{\Delta r_\perp}{\lambda}\right), \tag{3.13}$$

The angle θ shown in Figure 3.7 measures the angle between the deflection and the nematic director which is parallel to the backbone of the polymer. Combining equations 3.12 and 3.13, the value for λ is,

$$\lambda = \frac{1}{3}(1-S)l_p \tag{3.14}$$

The resulting mobility is dependent on $\langle N(\tau)\rangle$, a measure of the length and the molecular weight of the polymer, the monomer length a, and the persistence length l_p, The perpendicular mobility decreases with the nematic order S,

$$\mu_\perp = \frac{e\langle R_\perp^2(\tau)\rangle}{2\tau k_B T} = \frac{e\langle \Delta N(\tau)\rangle a l_p(1-S)^2}{9 k_B T \tau}. \tag{3.15}$$

LIQUID CRYSTALS

Small molecules depend of their π–π bonds to determine the distribution of their mobility. Liquid crystals polymers classify with these materials even though they are not small molecules. The charge transport for liquid crystals occurs through interchain transport, similar to that shown in Figure 3.7. Liquid crystals were thought to have charge transport that was mostly ionic. However, in the 1990s, the electronic carrier mobilities were determined to be in the order of 10^{-3} cm^2/V s. The contribution of the ionic transport depends on the viscosity of the phase of the material. The smectic and discotic phases of the liquid crystals are more viscous and therefore are less sensitive to ionic transport. This is also true of the polymeric phases of liquid crystals.

Polymerizable liquid crystals consist of a reactive end group decoupled from the conjugated cores by an aliphatic spacer. They maintain the charge transport of the self-assembled morphology when polymerized within the mesophase. The reactive end groups affect the charge transport, resulting in a reduction in the mobility that varies from a factor of four to two orders of magnitude.

Most liquid crystal molecules consist of a π-conjugated moiety and surrounding hydrocarbon chains. It is the π-conjugated moiety where the charge stays locally during the carrier transport events. The structure determines the type of carrier transport, whether it is one-dimensional, two-dimensional, or three-dimensional. The nematics do not have positional order. Thermal fluctuations will have the charge moving from one molecule to another in a three-dimensional manner. The carrier transport is thus considered to be three-dimensional. In smectics, the thermal fluctuations will have the charge moving from one layer to the next layer, in a two-dimensional hopping conduction. The discotics organize in a columnar fashion, interrupted by defects that depend on the size of the π-moiety groups. Therefore, the carrier conduction is one-dimensional in this case. The transport in smectics and discotics is described in more detail below.

The mobility in liquid crystals still depends on the Gaussian distribution due to the variation in energies and distances in many organic semiconductors. The distribution depends on the phases considered. The dependence of the mobility on the energy and the electric field for the liquid crystals is expressed by a relation that is very similar to that found for the polymers in equation 3.4,

$$\mu = \mu_o \exp\left(-\left(a\frac{\sigma_g}{k_B T}\right)^2\right) \exp\left(C\left(\left(\frac{\sigma_g}{k_B T}\right)^n - \Sigma\right)\sqrt{E}\right), \quad (3.16)$$

which is valid for the discotics as well as the calamitics. The constants a and n depend on the liquid crystal phase studied.

Discotic liquid crystals have the liquid crystal director perpendicular to the discotic core that contains in general the aromatic cores. The motion of charges is proportional to the mean-square distance traveled. Assuming the direction of the director is the z-direction, the mean-square distance corresponds to the z-axis, $\langle z^2 \rangle^{1/2}$ (see Figure 3.8). The charges will travel in a 1D manner. This transport will be interrupted by the presence of defects as mentioned above. The charge transport will still occur in the case of parallel columns separated by an amorphous region in the isotropic system. The transport will depend on how large this isotropic region is, in a similar fashion to what was found for the percolation in the polymers, shown in Figure 3.6. The existence of these defects will be the limiting factor in the performance of a device that depends on the mobility and conductivity of the electrons and holes. Thus, the domain size ψ and the parallel correlation length $\xi_{\parallel}$, introduced in Chapter 2, which depend on the size of

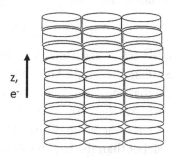

FIGURE 3.8 Sketch showing the direction the electrons will hop in a discotic liquid crystal. Since they run along the director's direction, they will travel in a one-dimensional manner.

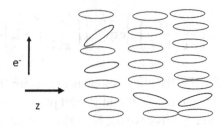

FIGURE 3.9 Electron transport shown for a calamitic (smectic) liquid crystal. It applies to the nematic also. It is in the direction perpendicular to the director, therefore it lies in the x–y plane. The charges will travel in a two-dimensional manner.

the amorphous region among others, limit the performance of the device.

The overlap of the π–π bonds is in the direction perpendicular to the director (Figure 3.9), for the calamitics. The hopping depends on how close the π–π bonds are. The dependence of the mobility is the opposite of that observed in the discotics. Perpendicular to the director, we obtain the highest mobility, $\mu_\perp$. If the direction of the director is the z-axis, the mean-square distances will be $\left\langle x^2 \right\rangle^{1/2}$ and $\left\langle y^2 \right\rangle^{1/2}$. The charges will travel in a 2D manner. The mobility and conductivity of the carriers will depend experimentally on the width or the FWHM of the perpendicular correlation length $\xi_\perp$, in addition to the domain size ψ.

EXCITONS

Excitons are generated when electrons in materials are excited with light or with a voltage, leaving a hole behind them. An exciton is a quasiparticle that consists of an electron and a hole held together by coulombic attraction. The excited electron has energy which is slightly less than the value of the LUMO and the hole has a value slightly higher than the HOMO for organic materials. Therefore, the exciton resides inside the energy gap. Two types of exciton exist in conjugated polymers: intrachain and interchain. These are shown in Figure 3.10. Intrachain excitons are formed

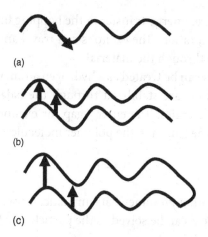

FIGURE 3.10 Types of excitons found in polymers: (a) intrachain; (b)–(c) interchain; (b) two chains close to each other; and (c) a chain the folds into itself.

by the extended π–conjugation found in sections of the polymer backbone. Interchain excitons are the result of chain segments coupling together because they are close to each other or because the chain folds back into itself. The intrachain excitons are responsible for the observed emission, whereas the interchain excitons do not emit.

Frenkel excitons are the most common excitons found in organic materials. The Coulomb interaction is relatively strong in the Frenkel excitons, with the size of the excitons in the order of the size of a unit cell or in the order of angstroms. Mott–Wannier excitons are weakly bound due to dielectric screening. Mott–Wannier excitons are more common in inorganic materials that have higher dielectric constants than the organic materials, with the exception of carbon nanotubes. Frenkel excitons, once formed, travel a small distance before they decay, due to the presence of defects or the lack of a well-established crystallinity. In organic materials, two models are used: the band or wave model for low temperature, high crystalline order, and the hopping model for higher temperature, low crystalline order, or amorphous state. Energy transfer in the hopping limit is identical with energy migration. The excitons can travel and carry energy but not charge through the material.

The excitons can be treated as a hydrogen atom, where the electron and the hole are stabilized through the balance of kinetic and potential energies. The energy gap for excitons depends on the inverse of the length of the polymer molecules:

$$\Delta E \sim 1/L. \tag{3.17}$$

This is easily, although somewhat simplistically, seen by assuming that this problem can be solved as the particle in a box of side L,

$$E_n = \frac{(h\pi)^2}{2m_e} \frac{n^2}{L^2} \tag{3.18}$$

The transition between level E_{n+1} and level E_n is given by,

$$\Delta E = E_{n+1} - E_n$$

$$= \frac{(h\pi)^2}{2m_e L^2} \left(n^2 + 2n + 1 - n^2 \right)$$

$$= \frac{(h\pi)^2}{4m_e N a^2} \tag{3.19}$$

where N is assumed to be n, and L=(2N−1)a, n is the top of the HOMO and n+1 is the bottom of the LUMO and in general N>>1. The longer the polymer is, the longer the wavelength of the light emitted if we are studying optical transition.

The decay of excitons is responsible for the behavior of a light-emitting diode (OLED). Photovoltaic applications require the dissociation of the exciton into an electron that goes into the LUMO band and a hole that goes into the HOMO band. To achieve this dissociation, either energy is required to overcome the binding energy or a trap formed by an impurity can be formed such that the exciton can reduce its energy by dissociating. In the absence of traps, an external source is necessary to separate the electron and the hole by a distance r_0 such that the thermal energy, $3k_BT/2$, is equal to their electrostatic energy, $e^2/4\pi\ \varepsilon\varepsilon_0 r$,

$$r_0 = \frac{e^2}{6\pi\varepsilon\varepsilon_0 k_B T} \qquad (3.20)$$

At the dielectric constants for organic materials and at room temperature, this distance has a value in the order of O (10 nm). If the distance is larger than r_0, they will become an independent electron and hole. An electric field would be a means of achieving this.

The electric field can be obtained by making a p-n junction, as shown in Figure 3.10, for an organic juncture. There is a drop in energy for both the electrons and the holes. The electron affinity as seen in Figure 3.11, rather than the position of the Fermi energy, determines which material is the donor (n) and which one is the acceptor (p). The excitation energy will correspond to the number of π-bonds in the organic material. If we use equation 3.7 for the

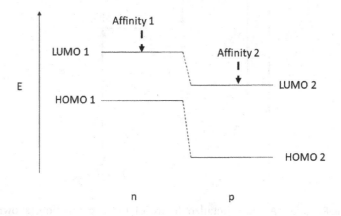

FIGURE 3.11   A simplified n-p juncture for two organic materials, showing how the electron affinity determines the band bending and which of the two materials acts as the donor and which one acts as the acceptor.

Marcus formalism for the rate of hopping, which we copy here, $\omega_o|H_{ij}|^2 \exp\left(-\dfrac{\Delta G - \lambda}{kT}\right)$, the Gibbs free energy ΔG is expressed in terms of the change in charge transfer, ΔG_{CT} and the ΔG of charge recombination, ΔG_{CR}, where,

$$\Delta G_{CT} + \Delta G_{CR} = -(\Delta E_{0-0} + E_b). \tag{3.21}$$

ΔE_{0-0} is the energy difference between the highest HOMO level to the lowest available excited state, designated in Figure 3.12 as S1, and E_b includes the coulomb attraction between the electron and hole, and is the difference between the optical and electronic bandgaps. These definitions are from the donor side since the initial excitation is typically in the donor. The charge-dissociation rate decreases with increasing disordering.

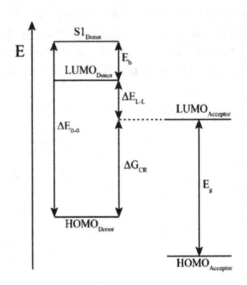

FIGURE 3.12 A more detailed figure of the n-p junction, showing the distance between the HOMO of the donor and the lowest available state in the LUMO of the donor. (Smith and Shuford, *Journal of Photochemistry & Photobiology A: Chemistry* 2018, **367**, 115–123.)

AGGREGATES

In semiconducting polymers and liquid crystals (both polymeric and small molecules), molecules align through their π–π stacking or through their dipole moments. This phenomenon exists in solution, especially for the polymers, where the molecules form aggregates. There are two types of aggregates. H-aggregates are those where the dipoles are parallel to each other (Figure 3.10b and c). They require an interchain geometry and do not emit in general. The J-aggregates are those where the dipoles align head to head or head to tail with respect to each other, making them intrachain aggregates (Figure 3.10a). They can also form between slipped π-stacked dipoles that belong to different chains, making them interchain aggregates.

We note that a lower energy corresponds to a longer molecule for polymers, if we compare the formation of J-aggregates to equation 3.19. It forms a longer wavelength exciton. The length produced by the H-aggregates corresponds to the molecule length and hence corresponds to higher energy. In general, interchain or H-aggregates will not be responsible for the observed photoemission, whereas the intrachain or J-aggregates will be. However, if we consider the J-aggregates produced by the slipped π-stacked dipoles that occur in the interchain direction, one can get the photoemission, especially if we consider small molecules. The distance and wavelength are longer than that produced by the H-aggregates.

COMPARISON BETWEEN MODELS

The efficiency of a photovoltaic depends in large part on the improvement of the order of the organic materials, at the mesoscopic and microscopic or crystalline level. This order maximizes the separation of the excitons at the donor–acceptor boundary and provides for an effective medium to transport the resulting electrons and/or holes to the corresponding electrodes. The Gaussian departure in the energy is a microscopic measurement that manifests itself in the ordering of the material. We also consider the

Lorenztian departure for the more disordered polymers and the liquid crystals. To obtain this departure or deviation, we need to use experimental techniques that look at the microscopic structure, such as, but not exclusively, X-ray or neutron diffraction. If we have a nematic field, we found in Chapter 2 that ψ and ξ are derived from the equation for the Gibb's free energy, G. $\langle \psi^2 \rangle$ is a measurement of the domain or grain size and $\langle \xi^2 \rangle^{1/2}$ measures how well the molecules are ordered within that grain size. $\langle \psi^2 \rangle$ and $\langle \xi^2 \rangle^{1/2}$ give Lorenztian relations that indicate the relative disorder of the nematic field. A similar suggestion is made for conjugated polymers, where the resulting equation is a Lorenztian. A Gaussian is the simplest distribution function obtained for the smectics and the discotics in general, and it will also give $\langle \psi^2 \rangle$ for the total intensity, which gives the grain size and $\langle \xi^2 \rangle^{1/2}$ for how well organized the molecules are in that grain.

All models associate the mobility of polymers and liquid crystals to the average distance in the direction of hopping. The mobility is modified by the presence of disordered sections in the material that control the percolation, determine the size of the grains in the material, and determine the order of the materials. The size of the grains determines the defects that reduce the mobility of the charges. This introduces a number of energy levels and distances in the semiconducting organic, characterized by a Gaussian distribution. The Gaussian departure in the energy is a microscopic measurement that manifests itself in the ordering of the material. We also consider the Lorenztian departure for the more disordered polymers and the liquid crystals. If we have a nematic field, we found in Chapter 2 that ψ and ξ are derived from the equation for G and therefore measure σ_g. A similar suggestion is made for conjugated polymers. If we have a well-ordered sample, we can compare directly,

$$\mu_\perp = \frac{e \langle R_\perp^2(\tau) \rangle}{2\tau k_B T} \sim \langle \psi^2(\xi_\perp) \rangle \qquad (3.22)$$

considering the full-width at half maximum $\xi_\perp$ for polymers,

$$\mu_\perp \sim \left\langle \psi^2\left(\xi_\parallel\right)\right\rangle, \tag{3.23}$$

considering $\xi_\parallel$ for the discotic liquid crystal, or,

$$\mu_\perp \sim \left\langle \psi^2\left(\xi_\perp\right)\right\rangle \tag{3.24}$$

considering $\xi_\perp$ for the calamitic liquid crystals. We use equations 2.23–2.26 from Chapter 2 to determine the values. Equations 2.23 and 2.24 from Chapter 2 are valid for the polymer. The value for the ψ's reflects the decrease in the value of μ due to the grain size and the defects in the material.

The change in the degree of ordering in the material can be obtained as a function of temperature T, a function of electric field, and a function of concentration when studying mixtures, among others. The mobility can be measured by time of flight measurements or indirectly by performing current–voltage measurements. These results can be related to the measurements of the order, which can be measured by neutron or X-ray scattering as mentioned before, and visualized by polarizing optical microscopy (POM), atomic force microscopy (AFM), transmission electron microscopy (TEM), and scanning electron microscopy (SEM).

REFERENCES

Adam, D., et al., *Phys. Rev. Lett.* 1993, **70**, 457–460.

Baldwin, R. J., et al., *J. Appl. Phys.* 2007, **101**, 023713.

Branch, J., et al., *J. Appl. Phys.* 2014, **115**, 164313.

Brown, A. R., *Sythetic Metals* 1994, **68**, 65–70.

Bushby, R. J. and Lozman, O. R., *Current Opinion in Solid State and Materials Science* 2002, **6**, 569–578.

Da Cunah, et al., *J. Mol. Model* 2019, **25**, 83. doi:10.1007/s00894-019-3953-3

DePodesta, M., *Understanding the Properties of Matter*, 2nd Edition, 2002, Taylor & Francis, New York..

Funahashi, M. and Hanna, J.-I., *Phys. Rev. Lett.* 1997, **78**, 2184–2187.

Geoghegan, M. and Hadziioannou, G., *Polymer Electronics*, 1st Edition, 2013, Oxford University Press, Oxford.

Hanna, J.-I., et al., *Thin Solid Films* 2014, **554**, 58–63.

Herbst, S., et al., *Nature Communications* 2018, **9**, 2646.

Hoffman, S. T., et al., *J. Am. Chem. Soc.* 2013, **135**, 1772–1782.

Iino, H., et al., *J. Appl. Phys.* 2006, **100**, 043716.

Kittel, C., *Introduction to Solid State Physics*, 8th Edition, 2005, John Wiley & Sons, Inc.

Kline, R. J., et al., *Macromolecules* 2005, **38**, 3312–3319.

Li, X., et al., *Phys. Rev. Lett.* 2012, **108**, 066601.

Lloyd, M. T., et al., *Materials Today* 2007, **10**, 34–41.

Ma, H. and Troisi, A., *Advanced Materials* 2014, **26**, 6163–6167.

Marcus, R. A., *J. Chem. Phys.* 1956, **24**, 966.

Martínez-Miranda, L. J., et al., *Appl. Phys. Lett.* 2010, **97**, 223301.

Miller, A. and Abraham, E., *Phys. Rev.* 1960, **120**, 745–755.

Noriega, R., et al., *Nat. Mater.* 2013, **12**, 1038–1044.

Rolland, et al., *Phys. Rev. Mater.* 2018, **2**, 045605.

Rothberg, L. J., et al., *Synthetic Metals* 1996, **80**, 41–58.

Smith, A. G. and Shuford, K. L. *Journal of Photochemistry & Photobiology A: Chemistry* 2018, **367**, 115–123.

Thompson, I. R., et al., *Phys. Rev. Mater.* 2018, **2**, 064601.

Zhang, W., et al., *ACS Cent. Sci.* 2018, **4**, 413–421.

CHAPTER 4

Liquid Crystals in Photovoltaics

I N CHAPTER 3, WE related the mobility of both non-liquid crystalline polymers and liquid crystals to the degree of order. We will present how liquid crystals used as a component of photovoltaics aid in their efficiency. Liquid crystals are used as a component of the photovoltaic, as a single molecule, as a chemically added end unit that converts a polymer into a polymer liquid crystal, as a cover for nanoparticles used in photovoltaics, or as a dopant. In all applications, the liquid crystals add to the efficiency of the photovoltaic.

Organic semiconductor molecules interact with weak forces that result in structural disorder that impacts the properties of the devices where these molecules are used. In general, the interactions that determine the ordering of liquid crystals and other anisotropic particles at the interface with another fluid have not been fully investigated. They depend on the details of the intermolecular interactions, be they attractive, repulsive, or a combination of the two. Increasing the interactions between the side of the anisotropic rod-like particles and spheres gives a preference for planar anchoring, while stronger interactions between the end of a rod give a preference for perpendicular alignment. In a phase-separated state, the interaction between the anisotropic rod-like particles and hard spheres favors a planar anchoring for repulsive interactions.

An organic photovoltaic consists of a polymer at least in one of the donor or acceptor units. The polymers have mechanical integrity that allows the photovoltaic to have flexibility without losing its shape. Liquid crystals that belong to the small molecule group do not have these desirable mechanical properties but have the ideal structural properties. Like all other small molecules, they have better electronic mobilities. Polymer liquid crystals do not have high mobilities as the small molecules, but have both the mechanical and the ideal structural properties that make organic photovoltaics more efficient.

We review the structure of the organic photovoltaic. One preferred structure is the dispersed heterojunction, shown in Figure 4.1a, as opposed to the bilayer structure, shown in Figure 4.1b. The dispersed heterojunction structure maximizes the interfacial area between the donor and the acceptor. It improves the

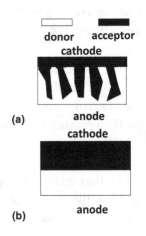

FIGURE 4.1   (a) A dispersed heterojunction versus; (b) a bilayer structure, separating the donor and the acceptor in a photovoltaic.

chances for the exciton to separate, and the electron and hole to diffuse to their respective electrodes. The molecular orientation at this interface determines how efficiently charge gets transferred, even though the long-range structure of this macrostructure is not ordered. This calls for a structure that consists of small crystallites that do not lead to the formation of molecular crystals or crystalline nanowires larger than the exciton diffusion lengths. This structure allows for both the electron and holes to move without scattering which would otherwise reduce the efficiency.

The photovoltaic has different processes that determine its efficiency: (a) exciton formation, which require the materials to be sensitive to the absorption of light from the solar spectrum; (b) exciton diffusion, which requires an almost free path from traps; (c) the dissociation of the excitons at the donor–acceptor interface; (d) the transport of holes and electrons to their respective electrodes, which again requires a free path, with a minimum of traps. This gives different efficiencies that contribute to the external quantum efficiency (EQE) depending on the probabilities that these four major processes occur. The process described in b can be subdivided into the probability of diffusion and the probability of dissociation, making the EQE a product of five probabilities,

$$EQE = \eta_{abs}\eta_{diff}\eta_{diss}\eta_{tr}\eta_{cc}, \quad (4.1)$$

where η_{abs} refers to the absorption, η_{diff} refers to the ability of the excitons to diffuse to the donor–acceptor surface, η_{diss} refers to the exciton dissociation yield, η_{tr} refers to the charge carrier transport throughout the device, and η_{ce} refers to the probability of the charge to be collected by the electrodes.

Let us look at the structure presented in Figure 4.1a again, to look at the resulting efficiency. The electron and hole mobility in an organic photovoltaic depends on how well the π–π bonds are stacked with respect to each other and how well these structures form percolating networks. This is illustrated in Figure 3.6 for a single polymer. As long as the second polymer fits such that it does not affect the percolation length of the first polymer greatly in a dispersed heterojunction made up of two polymers, it will aid in the transport of carriers. The concentration of the first polymer must be above the percolation length to achieve this condition. The interfacial region between the donor and acceptor regions can be conceived as consisting of a mixture that forms in the amorphous regions that exist in between the quasicrystalline regions, as shown in Figure 4.2. These quasicrystalline regions are formed by the overlap of π–π bonds. This structure facilitates the dissociation of the exciton and the mobility of electrons or holes. Outside the interfacial region, the quasicrystalline region will look like Figure 4.3 or Figure 3.6, and the percolation will determine how many carriers reach the corresponding electrode. The percolation is limited by the proximity of the lamellae to each other or the size of the disordered region that separates the quasicrystalline regions as mentioned in Chapter 3. The amount of carriers transported that arrive in the electrode improve the closer the electrode

FIGURE 4.2 Cartoon depiction of the region between the donor and acceptor regions in an organic or a hybrid photovoltaic. This region can consist of a quasicrystalline region of a polymer (upper drawing) or a percolating region (lower drawing) that facilitates the dissociation of the exciton.

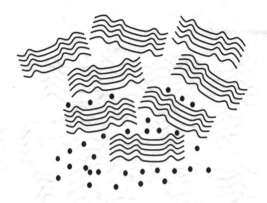

FIGURE 4.3 The region outside the interfacial region. In this region, the percolation determines how many carriers move to the electrodes.

is to the dispersed heterojunction donor–acceptor boundary since this increases the chances that the aligned nematic-type structure shown in both Figures 3.7 and 4.4 is near where the free carriers are generated. However, Figure 4.4 shows that this aligned structure is only a few layers thick at most and will not reach the boundary of the heterojunction, where the donor and acceptor meet. The straight line in Figure 4.4 shows the director of the nematic order the polymer adopts.

If a liquid crystal or liquid crystal polymer is substituted for the donor part, ideally the lamellae will be better ordered than in the polymer. The liquid crystal used will have π–π bonds stacked to form their structures. The acceptor will go in between the lamellae as shown in Figure 4.5. Their proximity will allow the dissociation of the excitons and the separation of the carriers. The carrier separation will be successful if the structure is as described above, that is, it does not equal the exciton diffusion length. Once they leave the interfacial region, the carriers will travel through an ideally ordered layer that will not have many traps or scattering sites. The ordered layer is determined by the substrate preparation. It extends beyond the first few layers from the substrate, which is the electrode, and into the boundary region, as shown in Figure 4.6. The straight line shows the director for the liquid

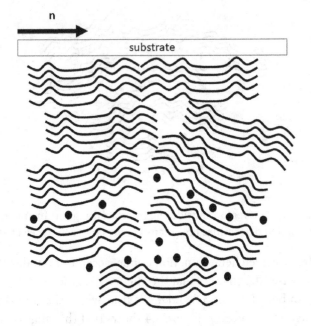

FIGURE 4.4 Sketch illustrating the juncture between the interfacial region and the substrate-induced quasi-ordered polymer. The sketch suggests that how well both couple to each other will increase the number of carriers reaching the electrodes.

crystal as imposed by the substrate. The position of the donor–acceptor boundary, although important, will affect the number of carriers less than for the polymer example as shown in Figure 4.4. The transfer of the carriers will depend on the structure more than on the percolation once they leave the interfacial region in this case.

Figures 4.4 and 4.6 show the relative distance from the donor–acceptor boundary where the ordered section of the non-liquid crystal polymer lies and the distance that the ordered part of the liquid crystal lies. The bulk structure of both materials is unaffected by the presence of the junction even when the aligned part of the polymer and the alignment of the liquid crystal depend on the substrate. If the appropriate type and phase of the liquid

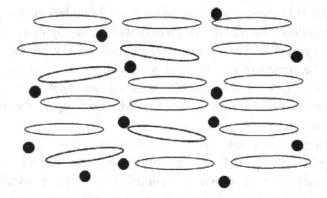

FIGURE 4.5 Sketch illustrating what happens when a liquid crystal or liquid crystal polymer is substituted for one of the regions in an organic or hybrid photovoltaic. The sketch illustrates the almost continuous region of order between the interfacial and the region outside the interfacial region.

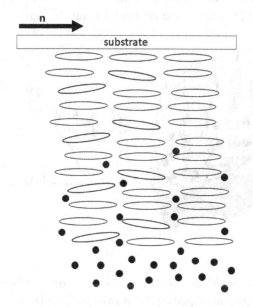

FIGURE 4.6 Sketch illustrating the possible continuity of the ordering from the substrate toward the interfacial region if a liquid crystal or liquid crystal polymer is substituted.

crystal is chosen, it will propagate to the boundary of the heterojunction. The transfer of carriers will be mostly determined by the percolation in the case of the non-liquid crystal polymer, and the structure will not play a significant role. The relative thickness from both of the substrates will determine the ability of the device to absorb the photon, which affects all the other probabilities.

The processes described in the previous two paragraphs will contribute to the EQC through the η_{diff}, η_{diss}, η_{tr}, and η_{cc} factors in equation 4.1. There will be a contribution to η_{diff}, especially if we have a liquid crystal structure. We look at the contributions to η_{diss}, η_{tr}, and η_{cc} first. Both systems will provide a path for the separation of the exciton in excess of 10–20 nm through the close mixture obtained. A side-chain polymer liquid crystal will provide a way for the acceptor to intercalate and form this close mixture, as seen theoretically in Figure 4.7a with its resulting scattering in Figure 4.7b, and an experimental result in Figure 4.8. η_{tr} will

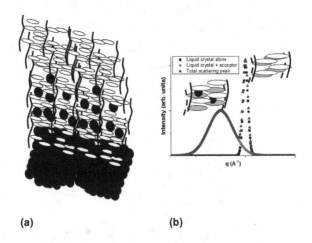

(a)　　　　　　　　(b)

FIGURE 4.7 (a) Model for the donor–acceptor interfacial structure shown for a polymer liquid crystal in an unpublished work of the author; (b) a theoretical X-ray scattering of this interfacial structure showing the interfacial nanocrystals and the crystals of liquid crystals. The polymer crystals are not illustrated in this sketch.

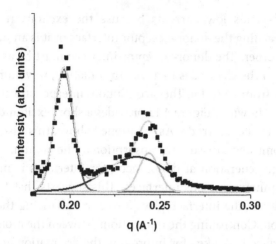

FIGURE 4.8 Unpublished glancing angle X-ray scattering (GISAXS) scan showing the interfacial structure on the right and one of the polymer peaks obtained for the polymer liquid crystal shown in Figure 4.7.

depend on the percolation in the first case and the structure in the second case presented in the previous paragraph. η_{cc} will depend partially on the ability of the carriers to come close by the electrodes. η_{diff} will depend on how well the liquid crystal or liquid crystal polymer is aligned. We examine how the liquid crystal structure aids in η_{diff}, η_{diss}, and η_{tr} in more detail.

DONOR–ACCEPTOR INTERFACE

The exciton diffusion length,

$$\left\langle R^2 \right\rangle = L_d^2 = 2D\tau, \qquad (4.2)$$

obtained using equation 3.9 in Chapter 3, is short for organics (about 5–10 nm). The structure of the donor–acceptor interface constitutes a bulk heterojunction consisting of photoactive layers or a single interface (Figure 4.1a). The exciton dissociation can only occur at this junction or interface; therefore, a large number of dispersion interfaces are preferred. The bulk heterojunction has a larger surface area where this dissociation occurs. The single interface yields low currents because the excitons recombine before reaching the donor–acceptor interface or it is an inefficient light absorber. The donor–acceptor interface must be as close as possible to the electrodes to avoid annihilation, in volume ratios that vary from 1:1 to 1:4. The bicontinuous interpenetrating nanostructure shown in Figure 4.1a provides a short exciton diffusion distance to the electrode. As mentioned above, this distance has to be monitored to assure the absorption of the photon.

Charge generation at donor–acceptor interfaces is more efficient at inhomogeneous interfaces. This suggests that the inhomogeneity of the interface plays a role in enhancing the charge separation. Controlling the interactions between the molecules at the interface is the key for improving the orientation and hence the charge transport. To transport charge, there must exist a percolated network that contains pathways along which charges can move easily. If the donor and acceptor are small molecules, they can form semiordered liquid crystalline phases if the small molecules are liquid crystals. They form motifs that reflect their anisotropic intermolecular interactions. These motifs can be pictured as those shown in Figure 4.7b, the top right drawing. The motifs cannot be much larger than the exciton diffusion lengths.

There exists a trade between obtaining an effective dissociation of excitons and providing a conductive path to the resulting free electrons and holes to their respective electrodes because of the shape of the interface. A control of the nanoscale microstructure that ensures both conditions is required of the blend, be it an organic–organic or in a hybrid blend as found in an organic–inorganic nanoparticle or nanorod blend. The ordered structure appears for these hybrid photovoltaics when the mixture is annealed at the liquid crystal temperature. The inorganic particle can be CdSe or ZnO or TiO_2, for example. The interface for this structure is presented theoretically in Figure 4.7a. The resulting X-ray scan is presented theoretically in Figure 4.7b, and experimentally, one X-ray result is presented in Figure 4.8. The result in Figure 4.8 shows a polymer peak to the left and the motif peak to the right.

The structure at the donor–acceptor interface influences the ability of an exciton to diffuse into this area and to be able to easily dissociate. The distance between the donor and acceptor must be within the diffusion length of the excitons, seen in equation 3.20 of Chapter 3 and determined to be between 10 and 20 nm. A structural solution that ensures a close donor–acceptor interface and works for organic-organic and hybrid photovoltaics is to have donor and acceptor molecules form a compatible structure in the donor–acceptor region where the possibility of separation is maximized. One way to attain this is to use copolymer molecules with side chains that are liquid crystalline in nature. This results in the optimum polymer-acceptor blending ratios. The size of the acceptor molecules and the spacing of the side chains of the polymer determine the formation of an intercalated nanostructure that will determine the performance of the device. The intercalated structure can be studied using X-ray scattering. The change in the position of the peak observed in Figure 4.7b or Figure 4.8 gives an idea of the size of the acceptor, whereas the full-width-half maximum gives an idea of the size of the donor–acceptor region. Similarly, a liquid crystal can be added to a copolymer to aid in the self-assembly of the polymer and an inorganic nanoparticle when the system is annealed at the mesophase temperature. At this temperature, a highly oriented nanoparticle region in the donor–acceptor region appears. Studies of the addition of a liquid crystal molecule show as the third unit in a photovoltaic that the liquid crystal can aid in adjusting the molecular arrangement and align the otherwise disordered polymer, and help to order it, increasing the efficiency of the photovoltaic.

Analysis using a combination of high-resolution TEM and polarization-dependent AFM shows that these intercalated structures generate a wire-type structure that covers the entire thickness of the film, except probably the region close to the electrodes. The liquid crystal arrangement of the donor is interspersed with this wire-type structure, forming disclinations at the points where both structures intersect. A sketch of such structure is shown in

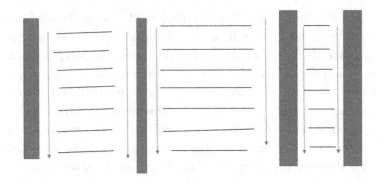

FIGURE 4.9 Sketch of the structure at the donor–acceptor interface. The lines denote the liquid crystal planes or the π–π planes. The solid rectangles denote the wire structure formed by the acceptor that is on average 20 nm. The arrows denote how the liquid crystal structure distorts close to the boundary of the wires.

Figure 4.9. They form ±1/2 disclinations apparently (Figures 1.12 and 1.13). Depending on the distance between these disclinations, the order parameter varies between 0.3 and close to 1. The disclination cores are coupled with pinholes observed with the AFM, although not every pinhole is associated with a disclination core.

The liquid crystal is partly responsible for the donor and acceptor proximity in a photovoltaic. This can be observed in the X-ray scattering study as the position of the peak and how close this is to the size of the acceptor. The liquid crystal is shown by the ellipsoids that hang from the polymer backbone in Figure 4.7b. They can be just single-molecule liquid crystals as shown in Figure 3.8 or liquid crystal polymers. We observe from Figure 4.7a that the acceptor, designated by the dark spheres, penetrates the liquid crystal polymer in the donor–acceptor region and forms the intercalated structure. This intercalated structure can be detected both in the perpendicular and parallel directions. The structure presented here is a hybrid photovoltaic, with TiO_2 nanoparticles about 3.4 nm in diameter as the acceptor.

CARRIER TRANSPORT THROUGH THE DEVICE

The process of recombination decreases the number of free charges available for transport through the device. Recombination occurs close to the donor–acceptor region. It is slowed down by the existence of traps that decrease the mobility by an amount given by equation 3.1, Chapter 3, and rewritten here,

$$\mu = \mu_o \exp\left(-\frac{\varepsilon_A}{k_B T}\right). \tag{4.4}$$

The traps affect the electrons more than they affect the holes as was mentioned in Chapter 3.

The tradeoff that exists between the effective dissociation of excitons that happens at the donor–acceptor interface and a conducting path of the resulting free electrons and holes to their respective electrodes that occurs through the rest of the photovoltaic device can be understood if we consider the structure. The conducting path contributes to η_{tr}. We consider first a semicrystalline polymer that does not have a liquid crystalline transition before crystallizing. The semicrystalline polymer is obtained by the formation of lamellae from the amorphous phase. The lamellae are formed by having chains fold over themselves or falling in order from different chains. This can occur from the nematic-type order obtained by some polymers because of the presence of substrates or the natural alignment between π–π regions presented in Chapter 3. They coexist with the amorphous phase that implies the existence of defects and grain boundaries. The temperature at which the semicrystalline structure forms depends on the size of the lamellae, with the shorter lamellae requiring a lower temperature. There are more π–π bonds as the chains increase, which increases the charge transport. However, the chains are not stacked for the entire length, as shown in Figure 3.6, Chapter 3, and they may not show a continuous structure to the structure sketched in Figure 4.9.

The advantage of the structure shown in Figures 4.7a is that the π–π bonds face the electrodes, which means that once the electron and the holes cross the donor–acceptor region, there is a straight path for the carriers to reach the electrodes. The holes travel through the liquid crystal or liquid crystal hanging units, whereas the electrons travel through the acceptor, be it an organic molecule for the organic-organic photovoltaic or through the nanoparticle or nanorod for the hybrid photovoltaic.

A liquid crystal structure has a one-dimensional or two-dimensional ordering that occurs at a higher temperature than the crystalline transition. A liquid crystal will have more π–π stackings tightly packed that span a large volume. In addition, the structure of the motif shown in Figures 4.7b shows that there is an expansion on the donor–acceptor region, which is observed in the accompanying X-ray scans of Figures 4.7b and 4.8. This structure may result in the appearance of grains when going from the donor–acceptor region into the donor region for the hole transport. The direction in which the carriers run in the liquid crystal is the more fluid direction that relaxes the appearance of the grains and traps. The structure sketched in Figure 4.9 has the advantage that the liquid crystal arrangement continues into the dislocation region produced by the wires and provides a direct path for the carriers.

The mobility of the liquid crystals is one of the highest of the organic molecules. Considering the relation of the mobility found in Chapter 3,

$$\mu_\perp = \frac{e\langle R_\perp^2(\tau)\rangle}{2\tau k_B T},\tag{4.5}$$

for both polymers and small molecules (liquid crystals) and its relation to the grain size

$$\mu_\perp \sim \langle \psi^2(\xi_\perp)\rangle.\tag{4.6}$$

A liquid crystal or liquid crystal polymer will provide a better path for carriers since this relation gives a path with less traps from the structural point of view. Considering only the shape of $\langle \psi^2(\xi_\perp) \rangle$, the shape obtained for the liquid crystal, even in the nematic is more well defined than the non-liquid crystalline polymers because of the disordered regions that generally exist between the ordered or lamellar regions in the polymer. Adding the energy dependence of the traps we can write the mobility and its relation to both the structure and the traps as

$$\mu_\perp \sim \langle \psi^2(\xi_\perp) \rangle \exp\left(-\frac{\varepsilon_A}{k_B T}\right). \tag{4.7}$$

In this expression $\langle \psi^2(\xi_\perp) \rangle$ and $\exp(-\varepsilon_A/k_B T)$ are not independent of each other, since $\langle \psi^2(\xi_\perp) \rangle$ absorbs the defects but not the traps that do not affect the structure directly, like ions. Equation 4.7 illustrates that the overall observed intensity is affected by all traps. Since the mobility and the conductivity are related, and the conductivity is the inverse of the resistivity, one can obtain graphs such as Figure 4.10, which shows the relation,

$$\psi^2(\xi_\perp) \sim \frac{1}{\rho} \sim \sigma \tag{4.8}$$

A better structure for organic acceptors can be obtained by getting π-conjugated structures that connect to the π-conjugated structures coming from the donor. For example, the π-conjugated copolymer structures arranged in alternating donor–acceptor structures that are used with some of the new non-fullerene acceptors. These include heteroaryl substituents that exhibit liquid crystalline behavior. Acceptor–Donor–Acceptor acceptors, with an extended and rotatable π-conjugation that results in higher electron mobility.

Hybrid photovoltaics have the advantage of a better mobility associated with the inorganic particles used as the acceptors.

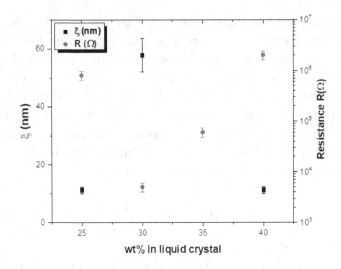

FIGURE 4.10   Graph showing that the resistance, which is proportional to the inverse of the conductivity, of the hybrid photovoltaic decreases as the parameter ξ increases. Graph shows that variation of the resistance and the parameter ξ as the concentration of the nanoparticle increases. (Based on Martínez-Miranda, APL (2010).)

They exhibit low mobility due to the organic ligand material (the functionalization compound) used to passivate the surface, and to the trap states that are the result of the same surface and the bonding to these materials. These functionalization compounds allow the inorganic compounds to mix with the organic compounds. Inorganic materials such as the nanomaterials used in hybrid photovoltaics can rely on functional molecules that can be mesomorphic to achieve different structures that can influence the charge transfer through them.

The mesomorphic materials can aid in the efficiency of the photovoltaic depending on how they are attached. There are a few solutions to reduce the effect of the insulating ligand. One of those replaced a ligand by washing the nanoparticle with hexanoic acid that resulted in a smaller distance between the nanoparticles and ensured a better charge transport. Other techniques involved repeated washings with methanol until a gel-like precipitate was obtained, or the treatment with oxygenated water, which leads to OH⁻ bonds that more easily attach through hydrogen bonding to the π-bonds in the donor molecules. Both these treatments lead to a better nanocomposite mixture and structure that lends itself to a better charge transport. A third approach is to choose a ligand that contains π-bonds itself and covers the particles such that the π-bonds are parallel to the particle. However, even though this approach aligns the particles and the molecules properly at the poles of the particle, they remain disordered around the equator of the particles. An important contributing factor is the alignment imposed by the donor–acceptor and the intercalation interface. This can induce structures with different interparticle separations that affect the charge transport.

Both the reduced conductivity due to a poor free carrier separation at the donor–acceptor region and poor charge transport in the rest of the device reflect in the series resistance measured by the ohmic part of the current–voltage curve. We show in Figure 4.10 that the value obtained for the resistance can be inversely related with ξ, and to ψ, as expressed in equation 4.8.

OTHER FACTORS THAT AFFECT THE EFFICIENCY OF THE PHOTOVOLTAIC

We have discussed how both the donor–acceptor interface and the transport of charges through the device and saw how these influence η_{diff}, η_{diss}, and η_{tr}. The remaining probability η_{cc} is related to how the electrodes are prepared and the materials used to aid in the orientation of the active materials in the photovoltaic.

We have already seen the effects of the electrode wall on semi-crystalline polymers, illustrated both in Figures 3.7 and 4.4. The alignments of the single-molecule liquid crystals used as a main member of photovoltaics devices are similar to what we find for the liquid crystal displays. Discotic liquid crystals will tend to align homeotropically with no additional or little substrate preparation, because of the geometry of the molecules. To align calamitic liquid crystals such that their π–π bonds are aligned along the electrodes, we can use the preparation used for displays. Before we look at the substrates, we review the difference of how the charges travel between the discotic and the calamitic liquid crystal. The charges travel in a one-dimensional (1D) manner for the discotics and a two-dimensional (2D) manner for the calamitics. Whereas aligning the discotics may be simpler than aligning the calamitics, the fact that the carriers travel in a 1D manner can represent a reduction of these carriers due to defects or grain boundaries that will act as traps. On the other hand, the carriers in the calamitics travel in a 2D manner, which offers an alternate direction for the carrier to take. Therefore, there is a trade between the alignment and the number of carriers through the liquid crystal.

The substrates can be prepared with different materials that aid in this alignment, although the less materials are used the less materials will influence the conduction of the device due to added traps. The alignment of liquid crystal polymers is similar to the single liquid crystals in general. Non-crystalline polymers can align in a nematic-like manner due to the associations of the π–π bonds. The electrodes are covered with a polymer cover, such as polyimide that facilitates the alignment and is generally insulating. The chemical properties of these materials and how they relate to the molecules in the donor and/or the acceptor of the photovoltaic determine how many carriers are allowed to go through.

OTHER DEVICES

The effects in the efficiency added by using liquid crystals or liquid crystal polymers are observed also in the perovskite photovoltaic, not considered in general as an organic-organic photovoltaic. Adding a liquid crystal or liquid crystal-like to the hole transporting material of the perovskite photovoltaic improves their efficiency, especially materials that exhibit discotic-type alignment. The efficiency of the perovskite is improved by incorporating a liquid crystalline polymer as a dopant to the structure. The dopant stabilizes the photovoltaic's behavior at room temperature. A study of the structure of the perovskite MaPbBr$_3$ shows structural disorder in the CH$_3$NH$_3^+$ ions while the PbBr$_3$ matrix is well defined. This suggests that an ion-exhibiting liquid crystal character may improve in the electron-hole separation and the electron and hole carrier transport and their ability of these to be collected in the electrodes. In other words, they affect η_{diss}, η_{tr}, and η_{ce}.

CLOSING REMARKS

Liquid crystals aid in the efficiency of organic photovoltaics and other hybrid photovoltaics. The interactions at the different interfaces and how they in turn interact with each other must be better understood to improve on the total efficiency. We must not only understand how the liquid crystals interact with either other liquid crystals, or with polymers or nanoparticles at the heterojunction, but how this interaction connects with the order imposed by the solid boundary of the electrodes to be able to improve on the efficiency. One consideration that must be taken into account is the distance between the inhomogeneous heterojunction and the substrate that imposes a nematic-like order in the polymers, and improves the order of the liquid crystals, be it a disc-like, or smectic or nematic order. The distance reflects the trade mentioned in the donor–acceptor interface that exists between an effective dissociation of excitons, providing a conductive path for the electrons and the hole to their respective electrodes. The distance also controls the absorption of the light. The liquid crystal part deals with the transportation of holes. Finally, the distance depends on the materials used, which points to the need of having a library of related liquid crystals to obtain the best material for the corresponding application that will aid in the corresponding computer simulations, which in turn will help in improving the effectiveness of the devices.

REFERENCES

Aldrich, T. J., et al., *Chem. Mater.* 2019, **31**, 4313–4321.
Bartlet, J. A., et al., *Adv. Energy Mater.* 2013, **3**, 364–374.
Boehm, B. J., et al., *J. Phys.: Condens. Matter.* 2019, **31**, 423001.
Branch, J., et al., *J. Appl. Phys.* 2014, **115**, 164313.
Bushby, R. J. and Lozman, O. R., *Current Opinion in Solid State and Materials Science* 2002, **6**, 569–578.
Cates, N. C., et al., *Nano Lett.* 200, **99**, 4153–4157.
Chen, W., et al., *Solar Energy Materials and Solar Cells* 2012, **96**, 266–275.
Draper, M., et al., *Adv. Funct. Mater.* 2011, **21**, 1260–1278.
Evans, P. E., et al., *J. Phys. Chem. C* 2018, **122**, 25506–25514.
Frost, J. M., et al., Nano Lett. 2014, **14**, 2584–2590.
Funahashi, M. and Hanna, J.-I., *Phys. Rev. Lett.* 1997, **78**, 2184–2187.
Han, G., et al., *Adv. Energy Mater.* 2018, **8**, 1702743.
Hanna, J.-I., et al., *Thin Solid Films* 2014, **554**, 58–63.
Jia, Z., et al., *Chemical Physics Letters* 2016, **661**, 119–124.
Li, F., et al., *Journal of Materials Chemistry* 2012, **22**, 6259–6257.
Lloyd, M. T., et al., *Materials Today* 2007, **10**, 34–41.
Ma, X., et al., *Journal of Materials Chemistry A*, 2017, **5**, 13145–13153.
Martínez-Miranda, L. J., *Appl. Phys. Lett.* 2010, **97**, 223301.
Martínez-Miranda, L. J., *Langmuir* 2016, **32**, 239–246.
Martínez-Miranda, L. J., et al., *Liq. Cryst.* 2017, **44**, 1549–1558.
Mayer, A. C., *Adv. Funct. Mater.* 2009, **19**, 1173–1179.
Meneses-Franco, A., et al., *J. Mater. Chem. C* 2015, **3**, 8566–8573.
Prehm, M., et al., *Chem. Eur. J.* 2018, **24**, 16072–16084.
Ramirez, I., et al., *Adv. Energy Mater.* 2018, **8**, 1703551.
Roders, M., *J. Phys. Chem. C* 2019, **123**, 27305–27316.
Shoaee, S., at al., *Adv. Energy Mater.* 2018, **8**, 1703355.
Sun, K., et al., *Natur. Commun.* 2015, **6**, 6013.
Takacs, C. J., et al., *ACS Nano* 2014, **8**, 8141–8151
Thomas, A., et al., ACS Energy Lett. 2018, **3**, 2368–2375.
Urbani, M., et al., *Chem. Soc. Rev.* 2019, **48**, 2738.
Veera, M. A., et al., *Adv. Energy Mater.* 2018, **8**, 1801637.
Wright, M. and Uddin, A., *Solar Energy Materials and Solar Cells* 2012, **107**, 87–111.
Yao, K., et al., *Organic Electronics* 2012, **13**, 1443–1455.
Yu, R., et al., *Adv. Energy Mater.* 2018, **8**, 1802131.
Yuan, K., et al., *J. Phys. Chem. C* 2012, **116**, 6332–6339.
Zhang, W., et al., *ACS Cent. Sci.* 2018, **4**, 413–421.
Zhao, F., et al., *Adv. Energy Mater.* 2018, **8**, 1703147.
Zhou, W, et al., *Adv. Energy Mater.* 2018, **8**, 1702512.

Index

Note: Page numbers in *italics* refer to figures.

Printed in the United States
by Baker & Taylor Publisher Services

Printed in the United States
by Baker & Taylor Publisher Services